GIS Simulation of the Earliest Hominid Colonisation of Eurasia

Kathryn Holmes

BAR International Series 1597
2007

Published in 2019 by
BAR Publishing, Oxford

BAR International Series 1597

GIS Simulation of the Earliest Hominid Colonisation of Eurasia

ISBN 9781407300139 paperback
ISBN 9781407330617 e-book

DOI https://doi.org/10.30861/9781407300139

A catalogue record for this book is available from the British Library

This book is available at www.barpublishing.com

BAR Publishing is the trading name of British Archaeological Reports (Oxford) Ltd. British Archaeological Reports was first incorporated in 1974 to publish the BAR Series, International and British. In 1992 Hadrian Books Ltd became part of the BAR group. This volume was originally published by Archaeopress in conjunction with British Archaeological Reports (Oxford) Ltd / Hadrian Books Ltd, the Series principal publisher, in 2007. This present volume is published by BAR Publishing, 2019.

BAR
PUBLISHING

BAR titles are available from:

BAR Publishing
122 Banbury Rd, Oxford, OX2 7BP, UK
EMAIL info@barpublishing.com
PHONE +44 (0)1865 310431
FAX +44 (0)1865 316916
www.barpublishing.com

Table of Contents

List of Tables

List of Illustrations

List of Appendices

Acknowledgements

This publication stems from my PhD thesis funded by the AHRB Competition B Postgraduate Studentships Fund, in the Department of Archaeology and Prehistory at the University of Sheffield.

I would like to thank my supervisor, Robin Dennell, for all his help and advice over the years. The germ of the idea for this thesis originated under his supervision at Undergraduate level and many illuminating discussions have followed. I would also like to thank my examiners, Andrew Chamberlain (Sheffield University) and John Gowlett (Liverpool University), for their helpful comments.

Various members of the GIS community have been extremely patient whilst I came to grips with the software and were happy to explain points many times; in particular Graeme Bonham–Carter (Geological Survey of Canada), Bahram Daneshfar (Ottawa University), Graham McElearney (Sheffield University), Gary Lee Raines (U.S. Geological Survey) and Steve Wise (Sheffield University). Assistance with the statistical analysis was provided by Nick Fieller (Sheffield University), who was extremely generous with his time.

I would also like to thank David Davison, editor at Archaeopress, British Archaeological Reports, for his advice in preparing this manuscript for publication.

Finally, although by no means least, I would like to express my thanks for the support that I have received from my family, particularly my husband, my friends and the staff and students comprising the research community in the Department of Archaeology and Prehistory at the University of Sheffield.

Abstract

In order to understand fully Plio-Pleistocene hominid migrations, a whole gamut of interacting factors must be examined. It has been hypothesised that the most important variables affecting hominid migrations out of Africa would have been environmental, since in most respects hominids would have acted in the same manner as any other large land mammal of the time. Therefore, as far as language barriers permitted, as many published sources as could be traced from palynological and palaeontological sites from the Late Pliocene and Early Pleistocene within east and northern Africa and Eurasia were included in this analysis. It has been shown that mostly as a result of politically-based reasons the data obtained was sparse in relation to the area covered, and is thus unlikely to be a true reflection of the fossil record.

By examining the relationship between hominids, fauna and vegetation at sites in East Africa, as well as at Dmanisi and 'Ubeidiya, it was possible to determine with what genera and in what environments hominids were preferentially associated. The use of a GIS simulation allowed the generation of a distribution map showing areas in Africa and Eurasia where the same combination of fauna and vegetation could be found. It was proposed that these areas represented areas to which hominids could have migrated during the Plio-Pleistocene. The accuracy of the model was tested and within the limits of the available data was found to be reasonably accurate.

Various other interdisciplinary aspects of Late Pliocene and Early Pleistocene geography were also examined. It was found that during the earliest part of the period in question the Northern Hemisphere environment was distinctly less arid than in modern times, with areas of desert substantially reduced leading to large expanses of grassland. With the onset of glaciation the climate became increasingly arid, with an environment close to the modern situation developing by the Middle Pleistocene. The effect of this on the timing and routing of hominid migration pathways was discussed.

Chapter 1

Introduction

The major issue to be explored in this publication is the migration of early hominids at and around the Plio-Pleistocene boundary. The use of Geographic Information System (GIS) software to predict and model the potential migration pathways represents the first use of this software for predictive reasons during this timeframe and will hopefully shed light on the many intriguing problems of this period.

Since the discovery of *Homo habilis* at Olduvai Gorge, Tanzania by the Leakeys in 1960 (Klein, 1989), it has been commonly believed that human origins lay within the 'Cradle of Mankind' in the Rift Valleys of Eastern Africa. Initial Australopithecine migrations expanded this centre in a southerly direction and westwards to Chad, although it would appear likely that it was not until the evolution of *Homo* that migrations out of Africa occurred. The timing of the first, and subsequent, waves of migration out of Africa is hotly debated, with estimates for the date of the initial migration differing by almost 1 million years.

Currently, there are few proponents for a migration date as late as 1 million years ago due to the ever increasing knowledge of hominids from Eurasia (e.g. Arribas and Palmqvist, 1999; Carbonell et al., 1995; Gabunia and Vekua, 1995; Gabunia et al., 2000a, b; Palmqvist, 1997; Parés and Pérez-González, 1995; Roe, 1995; Swisher, 1997; Swisher et al., 1994; Tchernov, 1987); however, the jury remains out on trying to put a more concrete fix on the date. The fossil remains and stone tools that have been recovered from sites such as Java, Dmanisi, 'Ubeidiya and Atapuerca are undoubtedly hominid or of hominid manufacture, but sediments suitable for dating have rarely been found in the vicinity. In cases where dating has been attempted, and frequently early or very early dates obtained, the accuracy of the dating techniques or the relation of the sediment dated to the fossil remains has been brought into question. Whilst there is hope that in the future a hominid site may be discovered where absolute dating techniques could be used with relative ease or that new methods could be developed to use on existing sites, there seems little way out of the current dating impasse. Despite these difficulties, many researchers believe that the Plio-Pleistocene boundary represents a useful starting point for debating migrations out of Africa (see papers within Straus and Bar-Yosef, 2001) due to the major environmental changes that occurred at that time.

This book aims to shed light on the 'Out of Africa' problem by approaching it from a different angle and proposing when during the Late Pliocene and Early Pleistocene hominid migration was most feasible and along what pathways the migrations may have occurred.

Whilst the exact mechanisms that caused early hominids to leave Africa are unknown (see Bar-Yosef and Belfer-Cohen, 2001 for a discussion), it seems likely that environmental factors would have been substantially more important than cultural ones. Specifically, hominid migration patterns during this timeframe should be investigated and interpreted no differently than the migration patterns of any other large mammalian land-dwelling species (Lewis, 1997; Turner, 1982). That is to say, environmental factors such as vegetation distribution and the location of food and water resources would have had the most impact on the distribution of early hominid species. It will be on the basis of this hypothesis that this study will predict and model early hominid migrations.

Fossil faunal evidence from Africa and Eurasia shows that there was much faunal interchange between the continents from the Miocene to Pleistocene periods. Most taxa appear to have migrated out of Africa, a path followed by hominids, and include deinotheres, hippopotamus, hyaenas and big cats including sabre-toothed, whilst the few taxa that migrated from Eurasia to Africa included pigs, horse and camel (Arribas and Palmqvist, 1999; Pickford, 1990). The fact that these migrations arose on an almost constant basis implies that barriers to migration between Africa and Eurasia were almost non-existent. It can thus be inferred that hominids could have left Africa almost 'at will' during this period.

Archaeological evidence from within Africa has shown that by the time of *Homo ergaster*/*H. erectus*, hominids were playing a far more active role in the shaping of the African environment. They had evolved from a predominantly vegetarian species through a scavenging phase into a species where meat was an important part of the diet. This had an impact on their interaction with herbivore genera and the carnivore guild, since they were likely to follow these taxa along their migration routes.

The migration of sabre-toothed cats, such as *Homotherium* and *Megantereon*, into Eurasia has been seen as intrinsic to the movement of hominids out of Africa (Arribas and Palmqvist, 1999; Lewis, 1997; Martínez Navarro and Palmqvist, 1996). The sectorial teeth of these cats resulted in their inability to consume large parts of their prey, leading to inferences that the landscape could well have been 'littered' with partially eaten carcasses. These would have represented ideal food resources for scavengers, including both hominids and hyaenas (Arribas and Palmqvist, 1999; Lewis, 1997; Martínez Navarro and Palmqvist, 1996; Turner, 1992), although the likely success of hominids competing with bone-crushing hyaenas has been debated (Lewis, 1997; Turner, 1992). It has been suggested that the most successful time for hominids in Europe would have been prior to 1.5 Ma when carnivores such as the sabre-toothed cats dominated, but the large hyaenas had not yet arrived (Turner, 1992). Bearing these points in mind, therefore, it seemed likely that investigating the Plio-Pleistocene faunal structure of Eurasia would assist with investigating hominid migration.

The use of GIS software to analyse and interpret the data places a more rigorous and empirical slant on the conclusions generated than hypotheses developed solely on the basis of observation. This particular investigation lends itself to GIS analysis since it has been hypothesised that environmental factors, particularly floral and faunal, would have had the most impact on migration patterns. Social factors can therefore be safely excluded, which benefits the model generation process since it has been shown that accurately modelling social behaviour can be difficult. This difficulty has led to its omission from many models, leading to charges of ecological fallacy and environmental determinism (Kamermans, 2000). In this instance, however, the environment is the primary agent of change and these accusations would be fallacious.

The main foreseeable problem with the research relates to the paucity of Plio-Pleistocene faunal sites from Eurasia, and so an important part of this study will be determining whether this lack of data is a true reflection of reality. If Eurasia was inhabited by few faunal genera during the Late Pliocene and Early Pleistocene, it makes it unlikely that early hominids would have been able to successfully migrate across the continent and reach Java in the Late Pliocene. On the other hand, if various factors have affected the recovery and/or publication of remains from Eurasia, then Plio-Pleistocene models of hominid migration can be more safely generated, with the caveat that up-to-date information should be added to the model as and when it becomes available.

The importance of this work will be in integrating data from various disciplines to allow a complete picture of Late Pliocene palaeogeography across North Africa and Eurasia. It will be crucial to understand fully the environmental factors that migrating hominids would have faced and how the movement would have varied from migrations in the modern world. The Late Pliocene climate, and consequently the geography, was very different from today's, and so a model based on modern-day climate and vegetation belts would inaccurately depict the pathways that the hominids could have taken. As a result, the GIS model will be based on Pliocene environmental reconstructions with the caution that such reconstructions, by their very nature, are limited and possess a certain amount of imprecision.

By assessing the association of hominids, fauna and vegetation in Africa and, through the use of the GIS model, plotting the same occurrence of fauna and vegetation in Eurasia, it is hoped that a distribution map can be produced indicating areas of likely hominid migration. The analysis of various forms of environmental data such as sea level changes, monsoon development, mountain orogeny and plate tectonics, to list but a few, will help indicate during what time interval(s) hominid migrations were most likely to happen.

Chapter 2

Introducing Geographical Information Systems (GIS)

2.1 An Introduction to GIS.

GIS systems produce complex computer models of the real world by integrating data in order to, amongst other things, solve problems, assist with decision making, model spatial behaviour and generate hypotheses (Heywood et al., 1998). GIS was first developed about 30 years ago by the Canadian Department of Forestry and Rural Development. The department required a database into which it would be possible to input large amounts of spatial data about the land and natural resources in Canada in order to determine whether any areas were under particular threat so that the land could be managed properly. It was understood that to try and collate all the data manually would take many years, and would then probably be out-of-date, and so the Canadian Geographic Information System (CGIS) was built (DeMers, 1997; Marble, 1990).

There is no agreed upon single definition of what comprises a GIS. It is more than a database into which information can be added, since questions can be asked of the database and new database layers created. It is certainly more than a computer aided design (CAD) package, although the graphical output produced is of an extremely high standard, as CAD packages are only capable of producing graphics (Savage, 1990). GIS are programmes that are able to analyse and manipulate spatial data (Marble, 1990); data can be stored, retrieved, updated and edited and the results displayed in either tabular, graphic or map form (DeMers, 1997; Savage, 1990) and queries can also be performed on the data resulting in the creation of new information (Johnston, 1998; Savage, 1990). That is, GIS are tools through which spatial data can be processed to allow better understanding of the relationship between the data and the space in which it is situated. This means that data about vegetation, settlements, transport networks etc. can all be incorporated into a single database for analysis, although each separate category of data would be represented on a different layer or coverage (Wise, 1998).

The acquisition of data suitable for entry into a GIS can be a very costly procedure both in terms of time and money (Wise, 1998). Some national bodies, such as the US Geological Survey and the UK Ordnance Survey, have already constructed digital maps of their countries from which paper maps can be produced and other data suitable for incorporation into a GIS is also available from the World Wide Web. Whilst in the United States much of the information held by the Geological Survey is freely accessible and can be downloaded from the Internet, the information held by the Ordnance Survey is sold at market price. Normally this amount can easily be met by utility companies planning the location of new services, but for the individual researcher the costs are usually prohibitive (Poos et al., 1995).

The information inputted into the GIS takes many forms and can come from various sources such as manually digitising or scanning existing paper maps, satellites, tabular data, aerial photography, digital elevation models, remote sensing and others (DeMers, 1997; Gillings et al., 1999; Heywood et al., 1998; Wise, 1998). This is an extremely tedious process that has been estimated to take between 50% and 80% of the time required for the whole GIS project (Heywood et al., 1998; Wise, 1998). The level of accuracy of data inputted into a GIS system needs to be high since when the data is analysed by any of the various techniques available within a standard GIS package, such as interpolation, creation of buffer zones and generation of slope information, any errors are compounded. Since most data sets are not error free, interpretation of the final output should include some recognition of the inherent inaccuracies (Heywood et al., 1998). It also needs to be born in mind that when dealing with historical problems ancient boundaries could well have varied from modern boundaries and so the use of pre-existing digital data may be inappropriate (Poos et al., 1995). Likewise, long-term GIS studies of past environments need to take account of changes in e.g. coastlines, river courses or vegetation. In these instances, the data has to be acquired from other, more wide ranging, sources; for instance ancient documents, archaeological research or palaeoanalysis.

GIS are capable of various analytical operations on both spatial and non-spatial data and on one or several data layers or coverages (Johnston, 1998). Data that matches a range of criteria can be selected by various means, e.g. through the use of tables or Boolean logic, so allowing further analysis of subsets of data. It is also possible to perform recalculations on individual coverages, for instance to convert altitudinal values given in feet to values in metres (Johnston, 1998). (Boolean logic is based on combining different variables with the expressions 'and, or, not, either/or' (Heywood et al., 1998: 107), and is used in many situations involving database queries where a specific part of the information held is required. Boolean logic can be carried out on GIS coverages through the use of techniques and algorithms built into the programme, such as intersect, overlay etc.).

Analysis of individual spatial layers is a task frequently undertaken with GIS packages and ranges from the generation of simple descriptive statistics, such as determining the most common vegetation type within an area, calculating distances or areas to more complex tasks involving the creation of buffer zones around specific features. It is also sometimes necessary to convert individual data layers between vector and raster data types to allow different types of analysis to be undertaken (see below for explanation of vector and raster data types) (Johnston, 1998).

In order for spatial information to be compared across different data layers, all layers must represent the same area, be of the same scale and projection, as well as be aligned in the same direction. Geometric transformation allows different layers to be matched correctly, and whilst this process is generally only undertaken on one layer at a time, another layer acts as a reference (Heywood et al., 1998; Johnston, 1998). One of the most common operations performed on multiple data coverages is the overlay operation that allows different data layers, of the same area, scale and alignment, to be compiled onto a single map (DeMers, 1997; Heywood et al., 1998; Johnston, 1998; Wise, 1998). Overlay procedures are vastly more complicated in vector data structure systems, since all the data lines have to meet and polygons have to be closed, and so require more computational power (see below). Furthermore, slight irregularities that occurred when digitising the original data layers can become visible during vector overlay procedures, perhaps resulting in errors in analysis and the drawing of incorrect conclusions (DeMers, 1997; Heywood et al., 1998). The overlay technique in raster data structures can be achieved through the use of map algebra. That is, the value of each cell in a data layer is added, divided, multiplied or subtracted by the value in the same cell in any number of other data layers so producing a new data layer (Heywood et al., 1998). It is also possible to create new data layers through the use of Boolean logic and the overlay technique, meaning that only cells conforming to certain criteria will be displayed in the new layer (DeMers, 1997). The overlay procedure in raster GIS is a powerful tool and far easier to implement than in vector data structures (Heywood et al., 1998). Clipping is another feature of GIS programmes that is carried out on multiple layers. In this procedure information that is outside the area of interest is removed allowing analysis purely of the area under consideration (Johnston, 1998).

Procedural studies are an important aspect of GIS research since they provide information about inaccuracies and errors that may be inherent within the various GIS programmes and how they may impact on the quality of data produced (Savage, 1990). There are many different GIS packages available, each of which uses slightly different mathematical algorithms when interpolating data to produce information on slope, aspect, contours etc, and as a result, each algorithm produces slightly different data. This means that despite entering identical data into various GIS packages, the use of different algorithms results in the generation of diverse data, all of which are equally 'accurate', and has the potential to result in the drawing of inaccurate conclusions (Savage, 1990).

Some of the advantages of a GIS are that the data files can be rapidly updated as new information is made available, since only the database requires the additional information and it is not necessary to redraw maps in order to assess the importance of the new material. GIS also allows information to be more easily shared than written records since it is a relatively simple matter to produce multiple maps or to send electronic information to other researchers (Allen et al., 1990). This is significant as it means that researchers working within the same area do not have to replicate each other's work.

Despite the fact that knowledge of these systems has been available for some time, it has only been recently that the technique has been applied to many disciplines including forestry, land registry, transport and facility planning, civil engineering, agriculture, environment, ecology and archaeology (Tomlinson, 1990).

2.2 Types of GIS Data Structures

Whilst all GIS programmes store data regarding the location and attributes of spatial features, there are two types of GIS in current operation, those using vector data structures and those using raster data structures (Johnston, 1998; Wise, 1998).

2.2.1 Vector Data Structure

Vector output is similar to that produced by traditional maps (DeMers, 1997); that is to say, the spatial information is inputted as points, lines or areas (polygons) (Johnston, 1998; Wise, 1998). Attribute data describing each feature is kept in a separate, but linked, database (DeMers, 1997; Gillings et al., 1999; Wise, 1998). In the simplest form of vector data structure the spatial locations of the features are identified through the use of x, y co-ordinate pairs. A single pair would identify a point, whilst the locations of larger features require many more points to define their shape; with lines represented as sequences of co-ordinate pairs and areas as interconnected lines with the first and last co-ordinate pairs being identical (DeMers, 1997; Heywood et al., 1998; Wise, 1998). The more pairs that are used, the more accurate the representation of the spatial feature will be, particularly at higher resolutions when greater detail can be observed, however, an exact reproduction of the real world will not be produced (DeMers, 1997; Heywood et al., 1998; Wise, 1998).

The scale at which the spatial data is captured has an important effect on how the data is portrayed within a vector data structure, since on a small-scale map a building would be depicted as a point, whereas on a larger scale the building would have an area. Likewise, roads and rivers would be depicted as lines on small-scale maps, but in reality possess an area (DeMers, 1997; Savage, 1990).

This type of crisp definition, diagnostic of vector systems, is not usually suitable within archaeology as it is not possible to say, for example, that a zone of influence stops at a set point, as boundaries are often very grey areas (Savage, 1990).

2.2.2 Raster Data Structure

The above problem is rectified to a certain extent by raster data models, although the precise location of data is lost through this method (DeMers, 1997). Each data layer is represented by a matrix of cells or pixels and a numerical value, Z, that is the attribute value of the pixel (Savage, 1990; Wise, 1998). The pixels are normally square and of equal area (Johnston, 1998) and cover the entire spatial area to be analysed (DeMers, 1997). The presence or absence of the item of interest is then recorded for each pixel, frequently in binary form, with the attribute data of each pixel intimately linked to the GIS display (Gillings et al., 1999).

When using a simple raster data structure, different features, such as elevation, land use, vegetation type, geology etc, have to be stored as separate layers called coverages (DeMers, 1997; Heywood et al., 1998; Wise, 1998). However, if each feature occupies a different location in space, it is possible to store the data on a single layer with a different numerical code assigned to each feature type. This occurs in soil distribution maps where multiple soil types can be defined or in digital elevation models (DEMs) where each cell could represent a different height value (Heywood et al., 1998).

To compare the relationships between variables, by 'stacking' or overlaying coverages, all pixels must be of the same size, in the same location and in the same orientation; that is the co-ordinates of all four corners of each cell must be known. Therefore, the information inputted into a raster GIS has to be registered to a co-ordinate system so that the same place on various layers can be identified (DeMers, 1997; Johnston, 1998; Wise, 1998). The most common form of co-ordinate system is the Cartesian co-ordinate system, where the location is recorded in terms of x, y co-ordinates, which in Great Britain is based on the National Grid (Wise, 1998).

Raster data is easy to manipulate mathematically using algebra in overlay or surface operations, since different layers can be combined using the values present in each pixel (DeMers, 1997; Savage, 1990). A drawback of this system, however, is that the graphical output can appear blocky, with boundaries showing as a step-like outline due to the discontinuous nature of the information (DeMers, 1997; Gillings et al., 1999; Johnston, 1998). The level of accuracy is also reliant on the size of the pixels. Large pixels could result in over generalisation of a phenomenon; that is making a vegetation type seem widespread whereas in reality it is extremely restricted in extent. On the other hand, small pixels cause a huge increase in the amount of data needing to be stored which is impractical for many situations (DeMers, 1997; Gillings et al., 1999; Savage, 1990). A compromise on the amount of resolution required therefore needs to be reached.

2.2.3 Vector/Raster Comparison

Both vector and raster GIS have strengths and weaknesses in different areas that can be broken down into four main groups: data storage requirements, accuracy by which they are able to reflect features in the real world, ability to change scale and analytical abilities (Heywood et al., 1998; Johnston, 1998; Wise, 1998). It is therefore important to choose the correct data structure for the spatial information and this is dependent on the discreteness of the data, the intended application and the source data (Johnston, 1998). As mentioned previously, vector data structures are most suitable for discrete data that can be represented accurately such as roads or stream networks. Raster data structures are generally appropriate for spatial entities with uncertain boundaries, such as archaeological sites, continuous data including elevations, for analyses that include spatial modelling and if the data is already in raster format as occurs in DEMs (Johnston, 1998; Wise, 1998).

Raster data structures tend to have larger data storage requirements as a value has to be recorded for every pixel in the image (DeMers, 1997; Heywood et al., 1998), however, the development of CD-Roms and zip drives has made this less of a factor when determining which data model to use.

A greater level of accuracy can be achieved through the use of vector data models since it is possible to record the exact location of every point or feature included within the database (Wise, 1998). This potential level of precision, however, is frequently compromised through the use of existing data sources, where accuracy levels have fallen due to human error etc. This is a factor that requires careful monitoring when using GIS programmes (DeMers, 1997). The size of the pixels used with raster data models is the limiting factor when comparing the real world with the generated image. The locations of individual features are frequently lost, unless the pixel size is extremely small, and the resultant image is generally less precise than that which could have been produced through the use of vector data models (Wise, 1998). In certain instances, however, this approximation of the real world is exactly what is required, especially in situations with continuously varying data (Heywood et al., 1998; Wise, 1998). This is because it allows a unique value to be assigned to each pixel and prevents assumptions based on data of misleading precision.

Changes in the scale of the image are most easily handled by vector data storage systems as the exact location of each feature is known by its x, y co-ordinates. The only major problem with scale changes seen in raster data storage is if the required resolution is below the size of the pixels. This is because the data would have originally been entered at a relatively coarse level of accuracy and so getting an increased level of precision is not possible (Heywood et al., 1998).

Vector and raster data storage systems are capable of different analytical tasks and so frequently the questions that are going to be asked of the data determine which system is used. As the location of features is more accurately stored in a vector data storage system, vector

GIS programmes are frequently used in mapping and network applications. Likewise, calculations performed on the data are more accurate when using vector storage systems, but the analysis tends to be more complicated (Heywood et al., 1998). Raster data storage systems can be analysed more easily using map algebra and Boolean logic, since different layers can be combined using the values present in each pixel (DeMers, 1997; Heywood et al., 1998; Savage, 1990). It is also frequently utilised in modelling applications as the pixel-held data can be easily subjected to statistical analysis (Kvamme, 1989). Whilst this may produce less accurate results, it is undoubtedly simpler to perform. Raster data storage systems are most suitable for environmental characteristics such as vegetation, soil type and even archaeological site boundaries, where the edges of a feature cannot be depicted by accurate lines (Savage, 1990).

One of the main advances in recent years has been the development of GIS packages that permit both vector and raster data structures. This has been important, as it has allowed the attributes of raster data to be stored in a separate database with many attributes therefore being linked to each pixel. This has consequently reduced the number of required coverages and so lessened the data storage requirements. It has also given greater flexibility in analysis without the necessity of utilising two separate GIS packages, since modules present within the programme convert the data between the two different types of data structure. Therefore the advantages of both systems are available to the user (DeMers, 1997; Johnston, 1998).

2.3 GIS and Archaeology

GIS is a relatively new innovation in archaeology, having only been used since the mid 1980s. Archaeologists have to deal with large amounts of information that is intimately linked to the spatial landscape (Green, 1990; Kvamme, 1989). This could include the location of prehistoric burial grounds, ancient pathways and settlements and, on a smaller scale, distribution of pottery or bone fragments within a single archaeological site; so the benefits of using a GIS in archaeological situations could be large. Even artefact analysis can benefit from the use of GIS as outcrops of the suitable resources used to make the artefact can be mapped and trading or exchange links postulated.

Archaeological sites in a region are not randomly located, but are frequently influenced by social issues as well as outside factors such as landform, soil and vegetation etc. It is possible to analyse the importance of these individual factors by combining a statistical programme with GIS in order to interpret the relationship between environmental factors and archaeology. For instance, are habitation sites always located in warm, sheltered positions? This could then lead to the development of site location models where areas with a high potential for archaeological activity could be identified. In this way a variety of different strategies can be developed and their usefulness tested through further research (Kvamme, 1989). However, when attempting large-scale modelling, the sparseness of archaeological sites within a landscape does present major problems due to the lack of data points from which to develop trends and maps. This problem is discussed more fully in Section 3.

2.4 Conclusions

In all instances of GIS use, the nature of the data inputted into the system needs to be remembered (Savage, 1990). Digitising data by hand can be an extremely inaccurate process and problems can occur when entering data collated by other individuals as a certain amount of subjectivity will have been introduced. In these instances, the data may not be accurate enough to support the conclusions generated (Savage, 1990). Cross-checking of the data is therefore an integral part of the use of any GIS system.

It is important to remember whilst using GIS that it is not an end in itself – the maps and other information produced need to be interpreted (Allen et al., 1990). The information resulting from a GIS analysis allows researchers to test theories and produce models (Savage, 1990; Stine and Decker, 1990). GIS has the potential to study relations between humans and the environment through time resulting in a modelling of spatial, temporal and cultural processes. These models are thus more in-depth than more simple predictive or correlation models, and can then be used to construct hypotheses (Savage, 1990).

Chapter 3

Predictive Site Modelling and GIS

As was described in the previous section, modelling and predicting site locations are important functions of many GIS projects. Archaeological site location models are mostly used for cultural resource management purposes, to predict potentially sensitive archaeological locations before development so that they can be avoided where possible (Dalla Bona, 2000; Duncan and Beckman, 2000; Kamermans, 2000; Kuiper and Wescott, 1999; Savage, 1990; Stančič and Veljianovski, 2000; Warren, 1990a; Wescott and Kuiper, 2000). Their use has expanded into other areas of archaeological spatial research allowing the formulation of hypotheses as to where archaeological sites are located and why. Whilst this type of analysis is not new, with researchers overlaying maps of different features on light tables – or even windows – for many years, the use of GIS has quickened this rather tedious procedure and increased its accuracy (Duncan and Beckman, 2000). One of the ways in which this has been achieved is a change in focus from the restricted area of a 'site' to a 'land parcel' (Warren and Asch, 2000). This could well be a reflection of the way in which raster GIS works, in that the survey area is divided into a grid with each square thus representing a land parcel and sites falling within these land parcels. It is the characteristics of these land parcels that are then examined for the presence or absence of sites, or other dependent variables, rather than the sites themselves (Warren and Asch, 2000). For ease of reference, however, each land parcel is still generally referred to as either a site or non-site locality.

3.1 Types of GIS Predictive Models

The aim of predictive locational models is to try to determine the position of a variable within the landscape in an unknown time or place, based on either key theories that are thought to affect the variable or empirical evidence. That is they can be either deductive or inductive (Kamermans, 2000; Kohler and Parker, 1986; Kuiper and Wescott, 1999; Warren, 1990a; Warren, 1990b; Warren and Asch, 2000). In both types of models, however, it has to be shown that the variables are linked in a cause and effect relationship (DeMers, 1997). Successful models are able to predict the location of the dependent variable with a greater degree of certainty than would have occurred by chance, that is their level of accuracy is over 50% (Warren and Asch, 2000). Frequently, however, a map of the study area is produced indicating areas of high, medium and low potential for the variable being considered (Kuiper and Wescott, 1999).

Most predictive models within archaeology have been developed in order to determine the location of prehistoric sites within a landscape. They can also be used to suggest the distribution or use of resources within an area, such as outcrops of certain lithic sources (Church et al., 2000). Within ecological studies, including conservation and evolutionary biology and wildlife management, the function of GIS models is not greatly different, except that the variable being predicted tends to be biological; e.g. the distribution or ecological preferences of certain species, rather than archaeological sites, or habitats may be investigated (Kidd and Ritchie, 2000; Stoms et al., 1992; Wright et al., 2000). What this does imply is that many of the techniques utilised by ecologists in various predictive modelling projects could be profitably adapted for archaeological data. This is especially useful since GIS has a longer history of use in ecology than archaeology and consequently the theories supporting the models are better developed (Church et al., 2000).

3.1.1 Deductive Models

The basis of deductive models is that theory, or known facts (Stoms et al., 1992), can explain the effect of a variable in a cause and effect relationship. In terms of archaeological site location, this means that theories of human behaviour, based on cultural or biological insights from previous archaeological or anthropological work, should be able to provide an indication of where on the landscape sites would be situated (Dalla Bona, 2000; Kamermans, 2000; Kohler and Parker, 1986; Kuiper and Wescott, 1999; Warren, 1990a).

For example, anthropological research has shown that in many primitive societies the easterly direction is considered to be auspicious, due to the rising of the sun, and buildings are frequently aligned in that direction (Parker Pearson and Richards, 1994). It may be concluded, therefore, that more archaeological sites would be located on east-facing slopes than on slopes orientated in other directions. The researcher developing the deductive model would then utilise this expert knowledge in the generation of the model. As deductive models do not permit testing of hypotheses, the validity of the model would then be tested through the use of expert knowledge. Since the models themselves are based on this same expert knowledge all the information presented is a result of tautology and the models produced can be of little use in prediction (Pereira and Itami, 1991). Also, the expert knowledge may be inaccurate if incorrect conclusions had been drawn from previous work. A case in point would be in Indonesian Timor where the Atoni are prohibited from orientating doors towards the east, since the sun is forbidden to enter buildings (Parker Pearson and Richards, 1994), and so east-facing slopes may be consciously avoided. Despite this apparent drawback, deductive models may, however, be of more use in explaining and understanding human behaviour than inductive models (Kohler and Parker, 1986) since they tend to focus on social variables, but they are little used in isolation within archaeology (Kamermans, 2000) due to their lack of empiricism.

3.1.2 Inductive or Empirical Models

Inductive models are more commonly used than deductive models in archaeology (Kuiper and Wescott, 1999), probably partly due to the often mistaken belief that the use of statistics makes a model inherently more accurate, and are based on observed patterns in data (Ebert, 2000; Kuiper and Wescott, 1999; Warren, 1990a). They are especially informative when the variable being studied is not well-known (Stoms et al., 1992). In prehistoric situations, the most important economic factors for humans were linked to the environment, which is why it is believed that examining environmental variables will provide some information about site locations (Kohler and Parker, 1986).

There are two key assumptions implicit in inductive models. Firstly, that all factors being equal, the independent variable, e.g. archaeological sites, will be randomly located over the survey area (Kohler and Parker, 1986). Secondly, as all environmental factors are not equal, since certain locales are 'better' than others, there will be a direct relationship between the dependent variable and the spatial location of independent ecological variables (Johnston, 1998; Kvamme, 1989; Warren, 1990b). That is, the location of the dependent variable, e.g. an archaeological site, is at least partially determined by, and so correlated with, favourable environmental factors, such as topographical setting (Kamermans, 2000; Kuiper and Wescott, 1999; Pereira and Itami, 1991; Warren, 1990b). These types of models require much survey work to accumulate large amounts of data about both the independent and dependent variables with the aim of producing an accurate prediction, and work by maximising the environmental differences between the locations where sites do and do not occur (Kvamme, 1989). The probability that a site occurs at a specific location is then determined by examining the environmental variables statistically shown to correlate with site location (Warren, 1990b).

When determining archaeological site location, some inductive models require the examination of either a portion of the region in question or an area exhibiting similar characteristics to the region to which the model would be applied. That is, empirical observations would be made with data collected from site locations as well as general environmental information. To a certain extent, the type of environmental data collected is determined through deductive reasoning; that is, the researcher would envisage that the aspect and slope of a locality would determine whether it was a suitable location for human activity and so this type of information would be collected. However, it is also clear that the type of data pre-existing in digital format would also affect what variables were included in any model. This data would then be entered into a site database and mapped to show the distribution of both the environmental variables and archaeological sites (Kuiper and Wescott, 1999; Warren, 1990a; Wright et al., 2000). In GIS terminology, these maps are known as primary coverages. Any interpolation or manipulation of the data to obtain information more directly relevant to the site, such as slope, aspect or distance to water, would result in the generation of secondary coverages. These coverages are subject to the limitations discussed in section 2.1, and as such some assessment of their accuracy should be made (Warren, 1990a). Analysis of the mapped data then allows the correlation of known site positions with the environmental evidence to predict the locations of sites in the rest of the region or the other regions possessing similar environments and variables (Kohler and Parker, 1986; Kuiper and Wescott, 1999; Warren, 1990a).

It is not sufficient to state that if most sites in a region are situated on east facing slopes that that characteristic was specifically chosen for and that a correlation exists, it has to be demonstrated that the characteristics typical of site locations are substantially different from the background environment (Stančič and Veljianovski, 2000). It may be that the majority of the region under analysis faces east and so it would be expected that most of the sites would also be on east facing slopes. GIS allows this to be investigated more rigorously by determining the aspect of all localities (Kohler and Parker, 1986; Kvamme, 1989). For example, if the archaeological sites were truly randomly located, it would be expected that if 30% of the land faced east, 30% of the known sites would face east as well. This expected frequency can be investigated with statistical tests such as Chi-squared (χ^2) or Attwell-Fletcher, with the differences between expected and observed site frequencies determining what factors are significantly correlated with site location. So, using the χ^2 test, if 3% of the surveyed region comprised evergreen forest and 30 sites were known in the surveyed region, the expected frequency of sites occurring in evergreen forest would be 0.9. Whether the expected frequency differed from the observed frequency and by how much would indicate whether the variable was useful for prediction. By only choosing the environmental variables that have a greater than expected frequency of archaeological sites, the environmental variables present at the site location should differ from the variables present at random localities distributed throughout the research area (Kamermans, 2000; Kohler and Parker, 1986; Kvamme, 1989; Pereira and Itami, 1991). The other environmental characteristics would then be excluded from the model (Stančič and Veljianovski, 2000). Therefore, the independent environmental variables used within a predictive model should be highly correlated with known site location (Church et al., 2000), although several factors may play a part as to why the sites are non-randomly in the environment (Kvamme, 1989).

Most inductive models use probability analysis that requires the collection of empirical data from locations with no archaeological sites, that is non-site data where the dependent variable of archaeological site is absent as opposed to present, as well as from the sites themselves. This control group would normally be developed by randomly selecting sites from within the study area that were available for occupation but avoided (Wright et al., 2000). The presence of two mutually exclusive groups, in terms of the dependent variable, allows the use of logistic regression analysis by which the probability of site presence or absence for each independent variable is calculated (Kohler and Parker, 1986; Kuiper and Wescott, 1999; Pereira and Itami, 1991; Savage, 1990; Warren, 1990a; Warren, 1990b). This type of analysis is used when the dependent variable, in this case site location, is a categorical measure that is either present or absent, whilst the independent variables can be of either nominal, ordinal, interval or ratio scales (Kohler and Parker, 1986; Warren, 1990a).

Although discriminant function analysis could also be used to develop site prediction models, it has been discovered that logistic regression is more reliable in circumstances where the independent variables do not have statistically normal distributions, a common feature of much archaeological data (Kohler and Parker, 1986; Pereira and Itami, 1991; Warren, 1990a). However, logistical analysis does require an adequate sample size, a condition that cannot always be met when developing archaeological models, especially when only existing data are to be used (Kuiper and Wescott, 1999; Wescott and Kuiper, 2000), since non-site localities are many times more frequent than sites (Warren, 1990a). Logistical analysis also makes the assumption that the sample of site locations used to define the model are representative of the study area as a whole (Savage, 1990), something that cannot always be confirmed when examining archaeological data.

Once all the conditions for using logistical analysis have been met, the next stage is to run the test on both the site and non-site data. The predictive power of each independent variable is thus ascertained and a formula produced which when applied to the various predictor variables would determine the probability of a site being located where that variable is found (Warren, 1990a). It is then prudent to assess the validity of the model (see below, Section 3.1.2.1).

Not all models require the addition of non-site localities to the dataset, although they still need site and environmental data, as stepwise multiple regression could be used instead of logistic regression in the generation of the model, although site location is still the dependent variable. This particular method was devised as it is often not possible to collect information from non-site locations (Savage, 1990). This statistical test still carries several other assumptions, however. In common with logistical regression, they are that the sites used in the generation of the model are an accurate reflection of the entire study region and that more sites will occur in localities with the same characteristics as in the original sample. The stepwise multiple regression technique is then applied to the data to determine what environmental characteristics are mostly highly correlated with site location (Savage, 1990).

The key to the success of archaeological site prediction models is that archaeological sites tend to reoccur in favourable environmental places and are not randomly situated in the environment and so the models are able to exploit the differences between site and non-site locations (Kuna, 2000; Warren, 1990b). Most archaeological models are inductive, although they are also likely to include a certain amount of deductive reasoning in terms of what independent environmental variables should be studied initially (Warren, 1990a).

3.1.2.1 Testing the Validity of the Model

The map outputs generated through GIS modelling procedures have a large visual impact on their audience. In general maps are accepted at face value with little thought given to their errors and limitations and so it is important to understand the implications and potential of the map and what it can show. As such, it is essential that mapped results are as precise as possible so that conclusions drawn from them are accurate (DeMers, 1997; Kvamme, 1989; Lock and Harris, 2000). Inductive predictive models should be thoroughly tested for validity before they are widely applied to archaeological data. There are several methods that can be utilised, although not all are rigorous, but frequently the choice of validity test is restricted by the quantity of data available resulting in less comprehensive testing than in an ideal situation (Wright et al., 2000).

The internal accuracy of the model can be tested through an examination of the gradient of the predicted site probability ($0 \leq p[site] \leq 1$). Moving along the gradient, as there is a decline in the accuracy of site predictions there should also be a concomitant rise in accuracy in the prediction of non-site localities (Warren and Asch, 2000).

The best test of the accuracy of the model is when data independent from that used to generate the model is used to test the model predictions; that is a testing sample is employed. If enough data has been collected, it is often possible to omit some randomly chosen data points from those used to create the model. Once the model has been completed it would then be applied to this separate data in order to determine the accuracy of the predictions. Any suitable data found subsequently to the model completion could also be used to test the model in the same way (Wright et al., 2000). This tests the model against known conditions and obviates the need for extra survey (DeMers, 1997).

With the cross-validation technique, each site or non-site locality within the dataset is omitted in turn and the model constructed around the rest of the site locations. The model is then used to predict the location of the missing data. Once this procedure has been undertaken for every data point, the mean percentage of points accurately predicted as representing either site or non-site locations would reflect the accuracy of the model (Wright et al., 2000).

It would also be possible to test the validity of the model by surveying previously unstudied regions that were predicted by the model to represent areas of high potential for the dependent variable being studied (Wescott and Kuiper, 2000; Wright et al., 2000). This could represent an extremely time consuming endeavour, however, especially if the region to which the model was applied was large. Nevertheless, if the model were developed in order to generate hypotheses over the most likely location of sites within an area, then undoubtedly the use of field survey would be the most appropriate way of testing both the model and the hypotheses.

The use of a training sample to test the validity of the model is the least rigorous method of determining accuracy. (A training sample refers to the data points used in the construction of the model). That is, the same information that was used to generate the model is then used to test its accuracy by recording whether the model is able to correctly assign each data point to either the site or non-site group (Warren and Asch, 2000). By its very nature, however, the results of this test are always overly optimistic (Duncan and Beckman, 2000), suggesting that the model achieves a greater degree of accuracy than is actually the case.

3.2 Problems Associated with Inductive Predictive Modelling in Archaeology

As was discussed in Section 2, the majority of errors enter a GIS database during the data entry stage and these errors tend to be propagated as the various coverages are subjected to further analysis. This can be especially serious if existing data is incorporated into a GIS model, since the researcher has no knowledge of the problems or shortcomings of the dataset, or if several secondary coverages have been generated from the same primary coverage (Hageman and Bennett, 2000; Warren, 1990b). This is likely to occur when using DEM data to derive relief, slope and aspect information, since elevation is a continuous variable that is measured at discrete points with the algorithms used to generate the intervening measurements based on the principles of spatial autocorrelation. (Spatial autocorrelation means that localities that are closer together are more likely to be similar than localities that are further apart (Johnston, 1998; Kohler and Parker, 1986)). Knowledge about the different interpolation techniques is therefore essential to understand how accurately an elevation coverage depicts the real-life situation, as the use of different algorithms will affect the data. This is especially true as there is normally more than one test available for each procedure within the GIS package and the researcher may not necessarily know which one is most suited to his/her data (Hageman and Bennett, 2000; Harris, 2000; Kohler and Parker, 1986; Warren, 1990b). The limitations that the model may experience due to data inaccuracies should be clearly set out so that they can be considered in the context of the conclusions and hypotheses generated by the model (Duncan and Beckman, 2000; Ebert, 2000).

The use of non-site locations in logistical regression predictive models is another area by which errors are frequently introduced into the model scenario. Environmental information, forming the independent variables, is not normally collected from locales that can be unambiguously identified as non-site localities. Rather, it is assumed that the information comes from areas with no archaeological sites (Duncan and Beckman, 2000; Savage, 1990). Since the entire point of developing the predictive model is to distinguish between site and non-site localities, researchers need to be 100% confident that none of the non-site localities used to develop a logistical regressive are actually archaeological sites (Savage, 1990).

Most predictive models operate on an extremely broad scale, often as a result of the limitations of the data available to the modeller. However, this has the effect of assuming that resources are distributed evenly over the entire area within each land parcel (Kohler and Parker, 1986). This is unlikely to be true (Church et al., 2000), especially in areas with dramatically changing terrain where variables such as elevation, aspect and slope would alter considerably. This is a weakness exhibited by many predictive models and one that receives little attention in discussions about their accuracy (Church et al., 2000).

The majority of predictive models developed for archaeological situations have used modern-day data to determine topographic characteristics, such as height, slope, distance to nearest water source etc, that are believed to have affected where prehistoric societies chose to locate their sites (Church et al., 2000; Kuiper and Wescott, 1999; Warren, 1990b). This presupposes that the modern situation is a suitable analogue for the past, a conclusion that in many cases is simply not true (Duncan and Beckman, 2000; Stančič and Veljianovski, 2000). In these cases it is questionable how accurately the predictive model is able to determine site location since the model is only displaying coincidences or correlations that are observable in the present day (Ebert, 2000) and the dynamic situation of the palaeoenvironment is frequently ignored (Church et al., 2000).

Even ignoring the above problem and concluding that modern environmental variables do accurately reflect the ancient situation, many of the environmental variables used within a prediction model could stand in for other decision variables. That is elevation may represent a proxy measure for either length of growing season, amount of summer rain or winter snow and also vegetation type. Therefore, this means that prediction models tend to be generally applicable rather than apply to specifics, as in different models of different situations the importance of elevation could mean different things. It is not the level of elevation itself that is determining where the sites are located and so what actually determined the site location is often unknown (Kamermans, 2000; Kohler and Parker, 1986). This represents one of the problems of using correlations between archaeological location and modern physical conditions in inductive predictive models; that is correlation, whilst implying, does not mean causality (Kamermans, 2000).

One of the main criticisms of the use of site predictive models in archaeology comes as a result of the post-processual mindset of many of its detractors. It is difficult to include most of the qualitative and social aspects of archaeology within a GIS, although somewhat easier when developing deductive models, since how can one model the cognitive aspects of movement behaviour that are essentially unknown for most periods in the past? While different scenarios can be constructed, it is impossible to determine which is the most accurate (Harris, 2000). As a result, these types of variables tend to be omitted from inductive predictive models, leaving the researchers open to charges of ecological fallacy and environmental determinism (Kamermans, 2000). However, this is not due solely to the use of GIS technology, but rather a lack of accompanying theory as well as limited availability of relevant digital data. In order to solve this problem, researchers need to have a clear understanding of what environmental, social and cognitive or ideological factors affect the location of archaeological resources within a landscape (Kuna, 2000), something that requires much more work. The development of inductive models has been useful, however, as it has shown that there is an important gap in current understanding of many prehistoric societies.

Frequently all known archaeological sites in an area are used to try and determine correlations between the site localities and environmental factors (Kamermans, 2000). It has been suggested that this focus of prediction models on 'sites' does not truly reflect the way in which the landscape would have been utilised by people in the past and does not take into account the changing needs of different populations. Specifically, sites only form one component of a system that would also have included travel between different sites, and a knowledge of what activities needed to be undertaken either on the journey between the sites or at the new site would also have affected the choice of new site location. It should also be considered that many sites are reused, often for different purposes, and had different functions associated with different societies (Church et al., 2000; Ebert, 2000; Kamermans, 2000). As a result there is no reason to suppose that the same factors that led to the choice of location for a ceremonial hunter-gatherer feast would also apply to the siting of a Roman garrison fort. Therefore, where possible, site location models should be developed using only one type of site to generate the model; however, the sparseness of the archaeological record may make this unfeasible (Kamermans, 2000).

One of the most complicated aspects of predictive modelling is that it is difficult to model movement across a landscape; however, it ought to be attempted since just discussing movement in terms of distances is too great a simplification (Harris, 2000). It is possible to construct friction or cost surface coverages within a GIS, whereby travelling downhill is easier than uphill and expanses of water have to be avoided etc, but frequently the land is treated as an isotropic surface due to a lack of adequate information (Harris, 2000). This means that a site prediction model could predict the occurrence of archaeological locations in places where the energetic costs of reaching the place far outweigh any potential benefits of the site.

Some inductive modellers believe that their job is completed once a pattern or correlation has been defined between the dependent and independent variables, e.g. that archaeological sites in an area would normally be found on western facing slopes not exceeding 5 degrees and within 500 metres of water. There is no indication that this hypothesis may require testing and as a result the model is purely descriptive with no suggestions expounded as to why the typical situation for site location developed (Church et al., 2000; Dalla Bona, 2000; Ebert, 2000; Warren, 1990a). That is, inductive modelling frequently does not take account of what is causing the pattern, just observes that it is present (Kamermans, 2000). A more explanatory framework, including the effect of social variables, needs to be developed in order to explore the full potential of predictive site location models (Church et al., 2000; Kamermans, 2000).

3.3 Conclusions

Despite the disparity of use of inductive and deductive models in archaeology, the two types should not be seen as completely separate when undertaking archaeological site prediction since the best models will contain aspects of both (Ebert, 2000) so allowing a combination of both environmental and social variables. Modelling is useful as it can provide new ways of thinking about and understanding research problems. However, by undertaking analytical modelling the model becomes a representation of the real world and is also susceptible to the inherent assumptions present within the modelling technology that may not always be made explicit (Lock, 2000). The result of the model is not the end of the modelling procedure, but something from which new ideas and theories can be generated (Harris, 2000; Lock, 2000).

As a result of the problems and drawbacks of most existing predictive models outlined above, this research has set out to explain explicitly the mechanisms through which the data used in the generation of the model was obtained and the drawbacks associated with said data. It is only as a result of such clarity that the benefits of any predictive site model can be appreciated.

Chapter 4

Limitations of the Dataset

The data set used in this analysis has been obtained from various sources relating to faunal and floral remains recovered from Pliocene and Early Pleistocene sites falling mostly within the Northern Hemisphere. This data was chosen based on the hypothesis that it would have been the location of these resources that would have determined the potential migration pathways for early hominids. Whilst some studies into hominid migrations have been undertaken by looking at the distribution of archaeological remains across the landscape (eg Gamble, 1993), utilising this information would not have been appropriate in this instance. This is because many archaeological discoveries across Eurasia have been the result of surface collections and consequently the age of the artefacts involved is unknown. Even in controlled excavation scenarios, the age of the site or tool assemblage is frequently uncertain. It would be unreasonable to include stone tools that could date from only a few hundred thousand years ago in a model designed to suggest possible hominid migration routes between two and one million years ago. All the data that has been included in this analysis has been securely dated to the Plio-Pleistocene period.

4.1 Geographical Location of Data Sites

The land area under consideration comprised the East African Rift Valleys, North Africa, the circum-Mediterranean area, the Levant, Southern Europe and a wide band across the Asian continent between 15° and 50° north. As Illustration 1 shows, the sites had an uneven distribution across this area, with very few from large parts of Asia and most falling in clusters in East Africa, Eastern China, the Himalayan foreland and circum-Mediterranean area. Several offshore localities where data has been obtained though drilling projects also fall within the study region, as is illustrated in Illustration 1.

4.2 Choice of Data

As far as language barriers permitted, as many published sources as could be traced from palynological and palaeontological sites from the Late Pliocene and Early Pleistocene within the study region were included in the database.

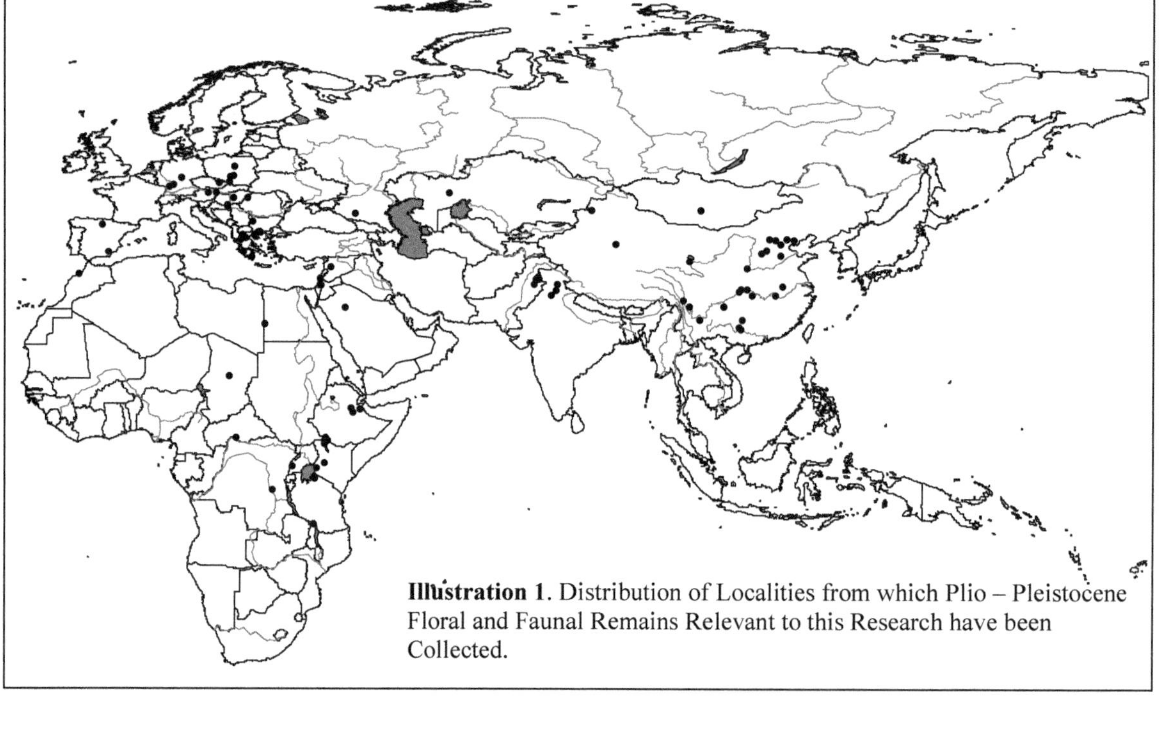

Illustration 1. Distribution of Localities from which Plio – Pleistocene Floral and Faunal Remains Relevant to this Research have been Collected.

4.3 Problems Associated With Obtaining Relevant Data

There was a massive amount of variation in the quantity and quality of data available from the countries within the study area as a result of historical and environmental factors as well as global and regional economic and political issues that have had different effects on the various countries.

4.3.1 Historical Background

Palaeolithic archaeology in Africa owes its roots to antiquarian interest during the latter half of the nineteenth century (Clark, 1997). This interest was aroused by colonial expansion within the continent and travellers' tales of ruins and rock art to be found in the interior. Stone artefacts were also found by explorers and were noted as being similar to those being recorded at the same time in France. Up until the 1920's, the archaeological discoveries were being made by geologists and naturalists with no expeditions mounted purely for their archaeological interest (Clark, 1997). Most discoveries were made by accident as Western geologists recorded the unknown areas. Once the richness of the fossil and archaeological remains became known, archaeologists accompanied the geological surveys and also organised their own expeditions to areas that geological work had previously highlighted.

The foundation of archaeological research in Asia was based on the hypothesis, argued by E. Haeckel, that human origins lay in Asia and was reinforced by the discoveries made in Java by E. Dubois (see Chapter 9). An archaeological expedition to Central Asia was also mounted by the American Museum of Natural History to look for hominids on the basis of H. Osbourn's belief that the climatic severity of the region would have provided an environmental impetus for hominid evolution (Pope, 1997). Whilst no hominid fossils were found there, the recovery of a tooth from Zoukoudian, and assigned hominid status by D. Black, led to excavation of the site funded by the Rockefeller Foundation. By the time of the Second World War Asia was more prolific than Africa in terms of hominid fossil material (Pope, 1997) and the future for archaeological research on that continent looked rosy.

Research on the two continents took markedly different paths after World War II. As is described more fully in Chapter 9, the exact provenance of the fossil hominid remains that had been recovered from Java was uncertain and doubts surfaced over the role that they may have played in human evolution. To compound this problem, evidence for an African ancestry for humans was recovered from Olduvai Gorge, Kenya and the focus for early hominid research gradually switched to Africa. This was partly due to the fact that funding was more easily obtained for excavation within Africa since it seemed more likely that hominid remains would be recovered from there but, as is described in more detail below, the political situation in many parts of Asia meant that work was unfeasible. This change in the balance of research work and the extensive history of geological work within Africa meant that there was a large discrepancy between the continents in terms of the amount of data applicable to this study and available for study.

4.3.2 Accessibility of Relevant Deposits

North African and Asian deposits of Late Pliocene and Early Pleistocene age have been less intensively studied than East African deposits, due partly to their relative inaccessibility as a result of overlying sediment (Brunet et al., 1995). The Sahara Desert covers over 3,500,000 square miles (9,065,000 square kilometres) (The Evergreen Project and Collaborative, 1998) and, except for areas close to oases or modern settlements, is generally unsuitable for conducting large-scale excavations. This is reflected at Bahr el Ghazal near Koro Toro in Chad (15° 58' N, 18° 52' E), one of two sites on the fringes of the modern Saharan region, where Pliocene sediments are slightly exposed (Brunet et al., 1995). Seventeen different localities have been discovered there recently, one of which contained fossil hominid as well as faunal remains, but seasonal weather constraints severely restrict fieldwork opportunities.

In 1963 the Yale Expedition undertook survey work at the Kurkur Oasis, which is in the Libyan Desert, Southern Egypt (23° 54' N, 32° 19' E) (Butzer et al., 1968). The duration of the fieldwork was restricted due to the limited supplies available to the team; however, they were able to take pollen samples from Late Pliocene and Early Pleistocene sediments. These early formations have been exposed due to erosion by the Wadi Kurkur, which now descends down a vegetated gorge and joins an aquifer. The sediments are thus visible on the wadi sides (Butzer et al., 1968).

Despite the number of geological surveys that have been undertaken in the study region, both in historical and recent periods, few local inhabitants were able to liase with foreign researchers about the Plio-Pleistocene rock formations or the material held within. This is because almost all geological work, particularly in Western Asia, has been undertaken for the petroleum and mineral industries by Western scientists. Obviously their work has concentrated on the oil containing sediments of a Carboniferous age, with Neogene deposits dismissed as 'surface', and so their data were not relevant to this study.

4.3.3 Climatic and Logistical Factors

Many researchers have been put off from working in most regions covered by this research as a result of inhospitable climate, although some do regard it as a challenge. Marked seasonal variations in both temperature and rainfall restrict the timescale of any excavation and long university vacations when lecturers are free from many teaching demands generally do not coincide with suitable weather for research.

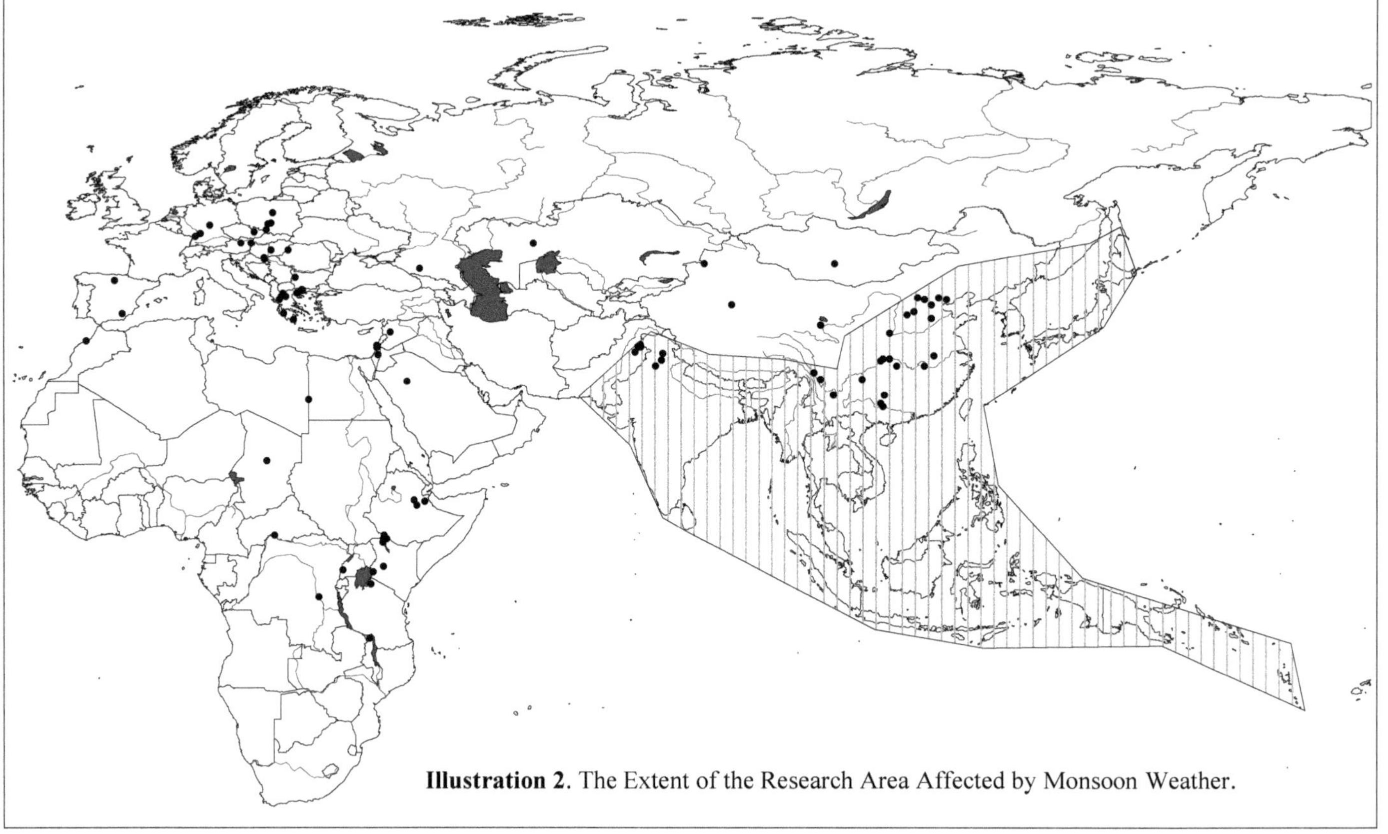

Illustration 2. The Extent of the Research Area Affected by Monsoon Weather.

The different environmental conditions that exist over Africa and Asia have played a large part in determining the quantity and quality of data available for study. A large proportion of the area covered by this study is subject to the effects of the monsoon; 'monsoon Asia' is crescent-shaped, reaching over 7,000 miles from Pakistan to Manchuria in northern China and covering approximately 6 million square miles (Cohen, 1972) (see Illustration 2). In this area seasonal rainfall is torrential and, combined with extremely high humidity levels, it not only provides an uncomfortable working environment during the Northern Hemisphere summer but also is not conducive to the survival and fossilisation of flora and fauna. The large rivers that cross substantial parts of eastern Asia deposit great quantities of sediment annually and impressive sedimentary sequences can develop, as illustrated by the Siwaliks in India and Pakistan. Any fossils that can be recovered, however, have normally been subjected to the effects of taphonomy resulting in incomplete assemblages that are often not in primary context and so environmental reconstructions are frequently impossible.

The African environment, particularly around lake shores, is more favourable towards preservation of ancient remains as sedimentation, whilst sometimes rapid, is generally gentle. Despite this advantage, much of Arabia and North Africa suffer from the same ecological problem, which is that the vast majority of the area is covered by desert and so provides an unwelcoming location for fieldwork. This is unfortunate, as the relative closeness of the coastlines of Somalia and Yemen could have resulted in migrations of animals, including hominids, across the Straits of Bab el Mandeb and into Asia.

The Pakistani government has limited access to archaeological areas by foreign field teams, with permits for each archaeological time period allocated to different countries. This means, for example, that only British run excavations are allowed to excavate sites from the Palaeolithic period (Dennell, pers. comm. 1999). As well as this, each country is only allowed to have one excavation team working at any time. However, some collaboration between Pakistani and foreign universities has been possible (Dennell, pers. comm. 2001). Nevertheless, this has restricted the amount of work that could be undertaken. It also means that researchers requiring information in order to further their own studies are reliant on the excavation team publishing their findings quickly, since the individual frequently may not be allowed access to the particular sites. Disputes between Pakistan and India over the State of Kashmir have made this region a 'no go area' and the unstable political situation in the region has not been helped by events connected to the September 11th 2001 bombing of the World Trade Centre in America. There is also a deteriorating law and order problem within Pakistan and difficulties due to corruption and inefficiency have made Westerners wary. These different factors have severely restricted access to parts of the important Siwalik sequences that contain long palaeontological and palynological records that would be applicable to this work.

More political, logistic-based problems affect foreigners wishing to conduct research in Indonesia since one of the major worries of working there is the appalling human rights record. Various monitoring agencies, such as Amnesty International, have reported incidents of torture, disappearance, racial and religious discrimination, a lack of freedom of speech and of the press as well as political intolerance (Frederick and Worden, 1992), and the people subjected to these atrocities included foreign visitors as well as Indonesian nationals. Despite these problems, during the 1980s researchers, mostly Dutch in collaboration with Indonesian nationals, undertook both reanalysis of the work conducted by Eugène Dubois and Ralph Von Koenigswald and instigated new excavations on Java in order to further understand the role that the islands played in human evolution (Bartstra, 1982; Braches and Shutler, 1984c; de Vos et al., 1982; Hadiwisastra, 1984; Hooijer and Kurtén, 1984; Pope, 1983; Shutler and Braches, 1985). Further work on dating of the sites was also carried out during the 1990s (Grün and Thorne, 1997; Swisher, 1997; Swisher et al., 1994; Swisher et al., 1996).

4.3.4 War and Peace

The archaeological record obtained from within Europe is without parallel due to the high calibre of researchers and the developed nature of the countries. Unfortunately, however, the period in question (Plio-Pleistocene) represented the very beginning of the last group of Ice Ages, and a time when much of Europe was uninhabitable due to the presence of glaciers. Indeed, it is commonly believed that most of Europe was not normally inhabited by hominids until only 500,000 years ago (Roebroeks and Kolfschoten, 1994). Southern Europe, particularly along the Mediterranean coast, would appear to have escaped from the worst of the climatic changes and archaeological excavations, complete with palaeontological and palynological work, that have taken place in France and Spain (Carbonell et al., 1995; Gibert et al., 1998; Lindsay, 1997; Lindsay et al., 1997; Martínez Navarro et al., 1997; Ollivier, 1986; Palmqvist, 1997; Parés and Pérez-González, 1995; Turner, 1992; Wolsan, 1993) have revealed evidence of human occupation as early as 800,000 b.p. (Carbonell et al., 1995) with some sites providing contentious ages of over 1 million years (Gibert et al., 1998; Roe, 1995). As a result, suitable data is available from numerous sites around the Mediterranean.

Sub-Saharan Africa, excluding Angola, Mozambique, Zaire and Zimbabwe (formerly S. Rhodesia) has also had a generally quiet post World War II period. Most countries, independent since 1960, have been extremely willing to accept assistance from their former colonial 'owners' in the guise of foreign workers assisting the state in numerous areas or foreign-funded research projects. These countries were interested in developing their archaeology since it provided them with knowledge of their history as well as a source of much needed foreign currency through tourism. Early prehistoric archaeology took off in this region with the discovery of *Homo habilis* at Olduvai Gorge, Tanzania by the Leakeys in 1960 (Klein, 1989), as before this date most research into human origins had been undertaken in Asia. Since then there has been almost constant research in both East and South Africa, resulting in a huge increase in the number of hominid remains recovered as well as a growth in knowledge of the local environment as a whole, and the type of habitats that early hominids preferentially inhabited, as well as the type of fauna associated with hominid remains (e.g. Behrensmeyer, 1978; Kappelman et al., 1997; Shipman and Harris, 1991; Spencer, 1997; Vrba et al., 1988–1989). This information has been recovered from pollen cores, fossil remains, taphonomic studies, functional analysis and comparative analysis (e.g. Behrensmeyer and Cooke, 1985; Bonnefille, 1983; Bonnefille, 1995; Fernández-Jalvo et al., 1998; Horowitz, 1989; Huntley, 1993; Plummer and Bishop, 1994; Vrba, 1980). As a result of the comprehensive coverage of the area a number of sites in the rift valleys of East Africa have provided data relevant to this topic; this is directly related to the amount of research that has been conducted in the region as a result of its importance in the early evolution of hominids. The ecology of the East African Rift Valleys during the Late Pliocene and Early Pleistocene is probably the best known in the world, and thus forms an important basis for this study.

China represents an unusual case as far as opportunities to conduct fieldwork in order to collate data are concerned, since for almost half of the 20th century the Communist ruling party closed the country to all foreigners. Access to China and data from palaeontological excavations is currently freely available, with, for example, a conference about human origins recently taking place in Peking and joint Sino-American field projects. Nevertheless, a major problem connected to the closure of China for 40 years is that the majority of publications are written in Chinese with few synopses of the evidence published in the Western world.

Disputes between Arabs and Israelis over the Palestine issue have been the cause of most of the violence seen in the Levantine area. Whilst various peace negotiations have been attempted since the creation of the Israeli State, the threat from terrorist activities remains high. Nonetheless, the data from Israel relevant to this study is of a quality seen only in Europe and East Africa, unfortunately, however, this does not extend into the neighbouring countries where little relevant research has been undertaken due to ongoing internal difficulties, such as the civil war in Lebanon. Terrorism has also represented a major threat to personal safety in other areas of the research area, particularly Egypt, and is another reason why relevant data is difficult to obtain.

4.3.5 United States and European Foreign Policy

After the Second World War, Soviet relations with the USA and other Western countries were extremely strained due to the Cold War and development and testing of atomic and nuclear weapons. Secret service agents from both sides of the divide were employed to enter the 'enemy's' country and determine state secrets. As a result, few western academics were allowed access to the region. It was not until Mikhail Gorbachev became general secretary in 1985 that a sense of openness developed between the East and West with relations proceeding on a friendlier basis. Data from the former USSR is even more limited due to the Communist policy of issuing passports to allow internal travel. Passports were only allocated to certain people (Zickel, 1989), and so many Russian nationals were unable even to conduct fieldwork within their own country.

The former USSR is still not completely safe for visitors or nationals. During the 1990s various republics such as Lithuania, Latvia, Byelorussia and Ukraine opted to leave the USSR and govern themselves. Whilst most of these separations were conducted peacefully, the wars in Chechnya, Georgia and Tajikistan and the fighting between Armenia and Azerbaijan represent considerable threats to security for visitors to these regions of the Soviet Republic.

During the 1960s and 1970s much archaeological exploration of Iran and Iraq was undertaken by Westerners with the local administrations happy to accept Western help. In Iran in 1979, however, the Pahlevi regime was overthrown and a revolutionary government came to power blaming the internal problems of the country on the shah's policy of 'modernising after the Western tradition' (Chapin Metz, 1987). As a result anything Western was shunned by the country leading to a restriction on all visas, essentially making it impossible for anyone from the West to visit the country (Chapin Metz, 1987). Whilst Iraq was still content to allow Western tourists and workers, its war with Iran from 1980 – 1988 invoked safety concerns in the West and in reality little research was carried out. The situation in the region was worsened by the invasion of Kuwait by Iraq resulting in the outbreak of the Gulf War.

Access to many of the other countries falling within the study area involves many personal safety concerns. Libya has managed to alienate itself from the Western world during recent years as a result of extreme terrorist activity. A British policewoman, Yvonne Fletcher, was shot by snipers hiding within the Libyan Embassy in London in 1984, as a result of which all diplomatic relations with Libya were severed. Libyan nationals are also believed to have been involved with the plane crash at Lockerbie, Scotland in 1989. Libya is also notorious for gun-running and so for security reasons the Foreign Office in Britain and national bureaux in other countries normally advise visitors to avoid the country. Afghanistan has been avoided by foreign visitors for a number of years as a result of war and civil unrest, and the situation in the region shows little sign of improving, although the recent overthrowing of the Taliban regime may cause matters to change. Also, the Union of Myanmar (Burma) has been declared 'off limits'.

4.3.6 Countrywide Factors

Many of the countries encompassed by this research have low GNPs (gross national product), and the allocation of this money in accordance with national priorities frequently leaves areas such as education and research drastically underfunded since money tends to be funnelled towards areas associated with national security or basic infrastructure. This obviously has a major impact on the number of nationals who are sufficiently well-educated to undertake research. Another factor affecting the quantity of information is that the level of public interest in scientific undertakings is often low in many regions of the study area. This is because in many countries the 'middle classes', who are normally most interested in scientific endeavours, represent a minuscule minority of the population, and are mostly in administrative positions, and many people have a day-to-day struggle to survive.

The majority of the countries within the research area suffer from poor education levels as a result of extreme levels of poverty. National dogma, be it Communism or devout religion, have also often taken precedence over educational needs. As a result of mediocre schooling, many local researchers have failed to meet the standards of academic rigour demanded by the Western world.

Most of the countries within the study area, except those falling within Europe where high schooling is normally maintained, have extremely low literacy rates and a generally poor educational standard that impacts heavily on the quantity and quality of research that can be produced. Whilst the vast majority of countries have made attempts to address this problem, governments are chiefly concerned with improving literacy levels, which are used to assess the stage of development of the country, rather than improving the quality of education.

Another factor affecting the standard of education and research in these countries is that throughout most of the 20th Century the majority of the countries in North Africa, the Middle East, South and South East Asia were considerably influenced by European powers. The object of most of these governments was to educate officials to serve the administration but not to create an unemployed élite class. As a result lawyers, doctors, teachers and government officers were produced, but there was much less success in educating scientists (Hourani, 1988).

As examples: the higher education system within Egypt that developed after the end of the Second World War could partly explain why there is a lack of data from this area of the world. Entry to university was originally determined on the basis of political beliefs with potential students only having to accept the ideology of socialism to be allowed entry, although this policy was later abandoned (Al-Farsy, 1986). This resulted in large numbers of graduates, some of whom did not possess the academic qualifications that would have been required in the West, who found it difficult to obtain work (Al-Farsy, 1986).

In Saudi Arabia the need to improve literacy and education levels has had to be combined by various governments with the studying of Islam, since most forms of education are not secular. Entry to higher education was based on academic qualifications, but even at this level of study religion remained a required element of the curriculum (Al-Farsy, 1986). A certain proportion of students have gone abroad to study at both the under- and post-graduate levels, where they were likely to receive a better education than if they had stayed at home; in 1984 13,000 Saudi's were studying mainly science-based courses abroad. This level is likely to have fallen, however, as scholarships provided by the government have been curtailed due to the number of places at university now available within the country (Al-Farsy, 1986). It remains to be seen whether the vast increase in student numbers results in an increase in the amount of research undertaken within the country, or if the system has expanded too quickly with lecturers unable to meet the high standards required in the Western world.

The Iranian education system prior to 1979 failed to provide an adequate level of teaching for most of the country's inhabitants. At the start of the century education was primarily for the children of the élite with few government funded schools. By 1940 the situation had improved, but it was not until the 1960s that the school system really expanded with around 75% of children attending primary school and about 50% going on to secondary school (Chapin Metz, 1987). However, there were few universities or colleges of higher education until the 1970s (Chapin Metz, 1987), when the expansion was funded through OPEC oil price rises. This poor level of education was exacerbated by the fact that many middle and upper class Iranian families sent their children abroad to study where they believed they would receive better education. As a result, there was no pressure to improve the system by the élite classes and many intellectuals chose to remain outside the country (Chapin Metz, 1987).

After the Revolution in 1979 the Iranian government acted quickly to reinforce Islam in schools with courses, textbooks and teachers that did not comply with the regulations barred (Chapin Metz, 1987). The universities also altered drastically after the Revolution; they became the focus of clashes in belief systems between students who advocated a more religious slant in terms of curricula and activities and those students who wished to maintain the existing secular system (Chapin Metz, 1987). This resulted in the government closing all higher education facilities between April 1980 and October 1983. Upon reopening, students and lecturers who were perceived to be lacking in faith were unable to return, leading to a large fall in the number of students (Chapin Metz, 1987).

In many respects the Iraqi education system resembles that of Iran. In 1921, when the modern state of Iraq was formed, education was limited to an élite class and the level of overall literacy was extremely low (Sousa, 1982). By 1958, most of the traditional schools had been replaced by public primary schools, but a large increase in population levels continued to keep literacy levels down. It was only after the revolution of 1958 that education was given any serious attention and attendance at primary school was made compulsory, but many non school age people remained illiterate (Sousa, 1982). Compulsory school attendance resulted in expansion of both the numbers of schools and universities, leading to an increase in the number of educated school leavers (Birks and Sinclair, 1982). A boom in the economy in the early 1970s, due to oil nationalisation and associated higher oil prices, provided the funds to plan a countrywide literacy scheme and laws were passed to ensure that the entire population attended classes on a regular basis (Sousa, 1982). Whilst this did improve educational standards within the country, it left many graduates unable to find work since few jobs were available at the upper end of the job market. As a result of this, in 1974 all unemployed graduates were taken into the public service (Penrose, 1979).

Education levels within Pakistan are extremely poor with the country possessing few schools and universities and the quality of teaching within most of these establishments has been a matter of concern (Blood, 1994). Whilst there has been an attempt to improve literacy amongst children, the adult education level is appallingly low. The main cause of this problem was the limited amount of resources that the government was prepared to spend on education. In 1990 3.4% of the gross national product (GNP) of Pakistan was spent on education, and this compared badly with the 33.9% of GNP that was spent on defence in 1993 (Blood, 1994).

In 1976 the central government of India assumed control of education for children up to fourteen years of age from local states, although the states still retained responsibility for setting standards in education (Heitzman and Worden, 1995). Five-year plans, which included educational policy and funding, were drafted by the central government, and in 1986 the government also drew up a long-term plan to guarantee that all children received similar levels of education with at least five years schooling by the age of eleven. However, these targets were not met within the approved timescale, partly due to the fact that, like Pakistan, India only spent 3% of GNP on education (Heitzman and Worden, 1995). Also, many schools were not brick built, with large numbers lacking drinking, toilet and library facilities and frequently only a single teacher was employed for the whole school. With children also allowed to work in cottage industries, it was not surprising that school was not an attractive proposition, resulting in a literacy rate of only 52.2% in 1991 (Heitzman and Worden, 1995).

There has been a large expansion in the number of students attending college or institutes of higher education since Indian independence. This has resulted in an increase in educational establishments, with universities, for instance, numbering six times more than at independence (Heitzman and Worden, 1995). The level of academic achievement necessary to enter college is not particularly high except in the areas of architecture, medicine, dentistry and business; even so, this has resulted in large numbers of drop-outs. To gain admission, however, the caste and regional background of the student were frequently examined; this was also true of students wishing to enter university in Pakistan. The government was the main employer of graduates, but a degree did not automatically ensure employment (Heitzman and Worden, 1995). All except national universities were subject to political intrigue since both the chancellor and vice-chancellor were normally representatives of the state government and the local government was also responsible for issuing grants (Heitzman and Worden, 1995).

In both India and Pakistan the quality of science education provided at universities since independence is not as high as could be hoped for, despite the benefits of foreign aid. Many first-rate scientists have moved to jobs within research facilities that are funded by the government and so offer more attractive wages and a substantial number have also moved to the West as a result of the limited opportunities in their own countries (Heitzman and Worden, 1995).

Education within the former Soviet Union was underpinned by a Marxist-Leninist ethos, that is the state wanted to influence the character of students in terms of morals and ethics as well as educating them (Zickel, 1989). Nearly all schools within the country were government run ensuring that they were overseen by the Communist Party of the Soviet Union (CPSU), which had overriding control over educational policy including content of the curriculum, methods of teaching and use of textbooks. The free education provided was intended to guarantee that future workers could provide the economy with qualified and skilled labour and to this end students were taught practical as well as intellectual topics. Graduates were supposed to pay back the government by working for two-to-three years in a job allocated to them (Zickel, 1989).

From 1970 it was compulsory for all students to complete their secondary education, and in 1987 more than 120 million people out of a total population of 282 million had completed both secondary and higher education (Zickel, 1989). Due to the massive differences in quality between rural and city schools, however, not everyone received the same standard of education. Schools also suffered from both shortages of space and equipment and had to run two, or even three, teaching shifts to ensure that all eligible children were taught. Most education specialists within the Soviet Union blamed these failures on the government only spending 4% of GNP on schools (Zickel, 1989). The curriculum in secondary schools included maths, science, history and literature, however, these subjects were frequently biased with only the official views taught, although under Gorbachev it was admitted that many history textbooks contained propaganda not facts (Zickel, 1989). All educational establishments had to teach within the boundaries of Communist dogma, but this was easily accomplished since most of the teachers were members of the Communist Party (Zickel, 1989).

In 1987 there existed 896 higher education establishments within the former USSR, but only 69 were universities with the remainder comprising medical schools, social science, natural science and engineering institutes, art academies, music colleges and polytechnics (Zickel, 1989). Entrance to university was academically challenging with knowledge of the area of study, Russian and a foreign language assessed, however, Communist Party backing added extra kudos to an application. It was also not unknown for places to be allocated on the basis of bribes or parental connections (Zickel, 1989). Theoretical features of arts and science subjects were taught at university, but only the development of knowledge and understanding were required, with analysis and evaluation of a topic excluded. Most researchers and teachers came from universities within Russia and textbooks in the natural sciences and humanities, for all institutions, were also produced by the Universities (Zickel, 1989).

The amount of graduate teaching undertaken in the Soviet Union was extremely small. Although all universities offered graduate courses, there were only around 100,000 graduate students in the mid 1980s with many undertaking their studies on a part-time basis whilst continuing to work (Zickel, 1989).

Education levels in China are higher than in most other countries within the study area, but the emphasis was generally on vocational skills such as engineering, science and technical training with the humanities more or less abandoned (Clubb, 1964). Under Chairman Mao the Communist Party was able to control education and research direction through numerous campaigns frequently associated with extreme punishment for non-compliance (Chang, 1993). A campaign was initiated in 1951 to 'persuade' the intellectuals to stop studying Western theories and instead follow ideas prevalent in the Soviet Union (Phillips, 1996). 1966 saw the start of the Cultural Revolution in which museums and books were destroyed and people seen wearing Western clothes, or even glasses, were attacked (Meisner, 1986). Mao believed that the dissident scholars had to be stopped in order to protect the Communist ethos; as a result, teachers were attacked by their pupils, writers and artists were victimised and other professionals attacked by violent youth gangs (Chang, 1993).

From before the start of the Cultural Revolution Chinese academics had been isolated from rest of the world (Goldman, 1981) with the Chinese Communist Party (CCP) controlling exactly who worked on what projects (Orleans, 1960). They also directed both the scope and focus of research as well as publication, particularly if it affected political and social beliefs (Charlton, 1997). The reforms of the Cultural Revolution led to the suspension of publication of academic, scientific and cultural journals (Meisner, 1986), further ensuring that the scholars were unable to continue their research or communicate with colleagues outside China. Between 1966 and Mao's death and the end of the Cultural Revolution in 1976, social science and humanities research ceased while science and technology institutions suffered from lack of equipment and fear of political attack.

Universities started to re-open in the 1970s, but students were assigned to courses by the Communist Party and so were not necessarily studying what they wished. Entrance to the Universities was not based on academic ability, but rather on the strength of the student's belief in the Communist system (Chang, 1993). Therefore, the abilities of students attending university were extremely varied. After their university course had ended, students were expected to return to their old jobs since Mao did not believe that education should change peoples' jobs as that would have been a bourgeois attitude (Chang, 1993). As a result there was little incentive to attend university.

Indonesia's government set themselves the goal in 1950 of providing universal education, in the form of six years of primary school teaching, but are finding this target hard to meet (Frederick and Worden, 1992). Suharto's orders in 1973 to allocate a portion of oil revenue to primary schools allowed nearly 40,000 schools to be repaired or constructed in the period to the late 1980s, which led to major improvements in the aim of supplying universal schooling. In 1991, however, the adult national literacy rate was 77%, although slightly higher for males, placing Indonesia in joint bottom place for literacy out of the members of ASEAN (Association of Southeast Asian Nations) (Frederick and Worden, 1992).

There has been a large growth in the number of higher education establishments in Indonesia with only ten institutions and 6,500 students in 1950 rising to 450 institutions and 237,000 students in 1970 and in 1990 there were 900 establishments and almost 4.5 million students. The number of students completing their studies on time, however, was extremely small (Frederick and Worden, 1992). Despite state universities being centrally funded they had more freedom in terms of curriculum than primary and secondary schools and the government subsidies meant that tuition fees were generally affordable. Lecturers' salaries at state universities were low when compared to a world-wide scale, although more than schoolteachers' wages, resulting in many lecturers holding additional jobs (Frederick and Worden, 1992). A large range of courses were offered at the universities, but around half the students studied social science subjects with 28% studying humanities and 21% specialising in science and technology in 1985. One of the major problems with the university education was that few professors held any sort of advanced degree; in 1985 only 4.5% of permanent faculty members in state universities held a doctorate and only a slightly higher percentage held Masters qualifications. This statistic has been hard to improve since there are few places in which to study for doctoral degrees within Indonesia and funds do not exist to allow students to study abroad (Frederick and Worden, 1992). Therefore, few Indonesians are qualified to conduct research into palynology and palaeontology.

Education levels within Israel are mostly of a Western standard, due in part to the massive immigration of Jews from Europe, America and Russia and the large amounts of money received from the United States in aid and donations. As a result, research and development activities within education are similar to those in developed countries (The State of Israel, 1999; Zahlan, 1988a). By the mid 1960s, 27,000 students were enrolled at universities and the country ranked 16th in the world in terms of research output, despite its minute size. In 1983 Israeli scientists produced 4661 publications from 292 institutions (Zahlan, 1988b). It is, therefore, hardly surprising that a large number of relevant publications regarding the 'Levantine Corridor' have been published by Israelis such as O. Bar-Yosef, E. Tchernov and A. Horowitz, since they were fully aware of the importance of this region in terms of hominid migrations out of Africa throughout the whole of human prehistory.

Due to the rapid growth in higher education that has been proceeding at an exponential rate as a result of the very low base level in all countries in the region except Israel, China and Russia, most countries, for instance Egypt and Turkey, rely on foreign study for postgraduate teaching. This means that postgraduates were not associating with lecturers actively researching their area of interest (Zahlan, 1988a) and so were falling further behind their Western counterparts. The number of students studying in higher education throughout North Africa and the Near East was not evenly split amongst the disciplines of the Humanities, Social Sciences and Applied and Pure Sciences. Information from the early 1980s indicates that by far the greatest proportion of students were studying Social Sciences, whilst the number of students reading Humanities based subjects was extremely low (Zahlan, 1988a).

Another factor limiting the number of people researching human origins has been religious extremism. Strictly speaking Islam and Christianity, amongst other religions, are fundamentally opposed to Darwin's theory of evolution, believing instead that God created man. In the more religious countries of this region, such as Iran and Saudi Arabia, secular views often bring severe retribution.

Taking these factors into account, it was hardly surprising to discover that most relevant research throughout the research area has been undertaken under the aegis of foreign palaeontological teams led by Western academics. Nevertheless excavations have been undertaken by nationals within the study area although, except in Israel and India, they are frequently regarded as somewhat flawed by Western academics. This is particularly true of China where, as a result of intellectual isolation since the Second World War, little notice has been taken of developments that have occurred in excavation techniques.

The publication of material also often reflects a certain naïveté in the handling of ideas and hypotheses. The material tends to be studied in an old-fashioned manner with a lack of critical treatment of older hypotheses and a failure to incorporate new ideas; again this was particularly prevalent in China, the former USSR and India. This may have been related to the cultural veneration of ancestors in these countries, but in fact means that there was a persistent belief that Eastern Asia was as important as East Africa in the evolution of hominids (Dennell, pers. comm., 1999). This has resulted in acceptance of many of the early dates proposed for sites within China, without a critical review of the evidence or a concerted effort to provide additional data, since many people are unaware of the problems associated with the material. Whilst this in no way implies that the dates from this region are unquestionably inaccurate, it does mean that extra care is needed when evaluating the sites and associated evidence.

In order to obtain theories and conclusions of a standard suitable for entry into Western hypotheses, it was frequently necessary to reassess many of the inferences made in excavation reports. Cases where this was not required have normally involved Western academic input into the publication, or have been produced by scholars from especially Israel, but also India and the former USSR, where academic rigour has usually been maintained.

In general there were few local researchers in most of the countries falling within the study area with whom Western researchers could collaborate. In the majority of cases research into human origins and evolution was not a high priority since there were many more pressing needs on which the country needed to spend income, such as housing, health and primary education. Where money was spent on developing the heritage of the country, the archaeology was dominated by research into 'civilisations' such as the Egyptian Pharaoh Dynasties, the Indus Valley (in Pakistan) and settlements in Mesopotamia (along the Tigris and Euphrates rivers). Exceptions to this can be found in India, Israel, China and former Soviet Union countries where, in certain instances, the impetus to research this particular area of archaeology was the mistaken belief in the early part of the last century of the importance of East Asia in the earliest evolution of hominids. In these countries archaeology is a well-developed discipline with a history of several decades, although political instability means that its long-term survival is not guaranteed.

4.4 Conclusions

A substantial problem relating to the lack of data from the Middle East, Western Asia and the Maghreb, was the attitude of many of the inhabitants to all things Western. Whilst education expanded to a remarkable extent within most of these countries, the graduates tended to look towards the state and bureaucracies to find work (Duignan and Gann, 1981). Chosen career paths were normally medicine, law, the military and computing since they offered both high status and high income. Scientific and technological research were looked down upon as part of the 'supposed "materialism," "heartlessness," and "irreligion" of the West' (Duignan and Gann, 1981: 43), although most countries were prepared to accept Western advances in areas such as medicine where the findings did not affect their orthodox beliefs (Dennell, pers. comm., 2000). In many countries within this region the universities were under strict government control, leading to only the 'official' version of facts being taught in many areas ranging from national history to foreign policy, economics and social affairs. In some cases boards of governors were in place to prevent the government from interfering in higher education, although governments normally kept close watch on student activities (Hourani, 1988).

Perhaps the most interesting feature to emerge from this study into the political and economic factors affecting the collection of data relevant to this study is the effect of the United States of America. American foreign policy has a major impact on what countries in the study area are considered safe for Westerners to visit and what Universities are prepared to have reciprocal relationships with Western institutions. There is also the apparent paradox that whilst America itself has no Palaeolithic and contains no early hominid remains, American researchers remain at the forefront of research in these areas.

Political problems of the later twentieth century are not the only reason why data was scarce over large regions of the research area. Palaeoanthropological research on the African and Asian continents developed in different ways and has had an impact on the quality and quantity of data that is currently available for study. Additionally, the differing environmental regimes have played their part resulting in differential preservation of floral and faunal remains across the research area.

Detailed studies of the political history of the last century, as well as other factors including weather conditions, has helped to explain why there were such large expanses of land, particularly in Asia, where no data was available for this analysis. Without a proper understanding of the limitations of obtaining data it was not possible to truly evaluate the evidence for Plio-Pleistocene faunal and floral occurrences across this vast area.

Chapter 5

Problems of using Faunal and Floral Data in Palaeoenvironmental Reconstructions

Floral and faunal data used to reconstruct palaeoenvironments are collected from both surface exposures and excavation. In an ideal world all floral and faunal remains would be preserved *in situ*, carefully protected in layers of sedimentary rock, waiting, as it were, to be found. Unfortunately, this is not normally the case since vast quantities of prehistoric time pass by without any record in the geological strata (White, 1995). In fact, it has been suggested that the geological record is less than 1% complete (Van Andel, 1981). Therefore, any remains used in reconstructions only represent a fraction of the possible diversity that would have been alive at any one time.

The level of accuracy that can be expected in attempting an environmental reconstruction on the basis of faunal and floral remains from the fossil record is further limited due to the effects of taphonomy. Without appropriate consideration of the type of processes to which the remains have been subjected, and how an assemblage may have been altered, any reconstruction will be biased (Andrews, 1992; Western, 1980). In order to interpret the changes accurately, the theory of uniformitarianism is relied upon; that is to say, the processes that are observable in the present also operated in the past. Whilst this may not be true in every case (for example, extinct animals may have acted in a way, or lived in an environment that has no modern analogue), it does provide a basis from which the remains may be analysed (Andrews, 1992).

Nevertheless, as a result of actualistic studies of modern day fauna, it has been discovered that it is possible to reconstruct environments on the basis of the animals found within them (Andrews, 1996). That is, under certain circumstances a faunal sample can contain enough information to enable one to reconstruct the original environment of the remains with reasonable accuracy. This technique is open to individual interpretation, however, and it is not unknown for various, equally plausible, scenarios to be proposed. Palynology is also a useful tool to employ when studying palaeoenvironments since identification of pollen grains and spores within a pollen core provides an indication of the type of environment that existed at the time the pollen was deposited (Scott, 1984). When used in tandem with faunal analysis the two techniques can act as self-checking mechanisms.

5.1 Taphonomic Considerations

It is through various taphonomic processes, the most important of which are fragmentation and disarticulation that occur as a result of the actions of humans, scavengers and weathering, that fossil assemblages become altered from the original living communities. Studying how the method of preservation affects the fossil record is an important aspect of taphonomy (Behrensmeyer et al., 1979; Behrensmeyer and Kidwell, 1985). Actualistic studies are frequently used in order to understand taphonomic processes, that is how assemblages in the modern world respond to taphonomic effects (e.g. Behrensmeyer's work in Amboseli National Park, Behrensmeyer and Dechant Boaz, 1980; Behrensmeyer et al., 1979). In this way the key diagnostic features that are observed, for example in bone collections produced by lions, hyenas or porcupines (Brain, 1981), or collections that have experienced allochthony (i.e. long-distance transport and sorting by fluvial action), can then be applied to fossil assemblages in order to come to a greater understanding of how a particular assemblage came into existence.

Animals do not necessarily live where their remains have been uncovered and transport of the body or skeleton after death could have occurred in several ways. Firstly, the animal may have died within its normal habitat but environmental factors, such as river transport, could have moved the animal carcass into a different environment (allochthony). The best way of determining whether an assemblage suffered from water transport is to examine the sorting of the bone remains since, in theory, fluvial sorting is extremely diagnostic. When disarticulated, bones such as vertebrae, sacra and podials are generally transported further than teeth or mandibles due to differences in density and shape (Behrensmeyer, 1975), however, articulated bones react in a more complex manner. Nevertheless, water transport does lead to disintegration of the original assemblage. Heavy bone parts transported by water tend to accumulate in slow moving sections of the river where there is insufficient power to move them further, leading to a deposit of unrelated material. The lighter material becomes abraded and is unlikely to be deposited and therefore will be missing from assemblages formed by water transport (Behrensmeyer, 1975).

It is not unknown for scavengers, including hyenas, to transport carcasses away from the area of death, perhaps to a lair (Behrensmeyer, 1975; Kruuk, 1972 cited in Behrensmeyer et al., 1979). This is unlikely to be a significant factor in bone transport from one habitat to another, based on observations in Amboseli National Park (Behrensmeyer et al., 1979), but it needs to be borne in mind since prey selection by carnivores could bias an assemblage. This is because most carnivore species hunt prey falling within certain size parameters (Andrews, 1996; Behrensmeyer and Dechant Boaz, 1980; Behrensmeyer and Kidwell, 1985).

'Time averaging' of assemblages represents another factor to be considered when using fossil remains to interpret past environments (Western, 1980). 'Time averaging' occurs when the remains from a faunal assemblage represent more than one segment of time, for instance the remains could be from a deposit that accumulated over several seasons or years, with no obvious stratigraphic zonation to indicate that time had passed. As a result, contemporaneity of remains found within the deposit cannot always be established. This may have the result of mixing fauna that did not coexist in the same setting, therefore giving a false environmental reconstruction (Behrensmeyer, 1985; Stern, 1993). 'Time averaging' can be useful, however, in indicating long-term patterns within a community (Behrensmeyer and Kidwell, 1985; Stern, 1993; Stern, 1994). The kind of assemblage that does not usually suffer from 'time averaging' is one that formed under catastrophic circumstances, for example in a flood. In certain circumstance, though, even these types of assemblages can be 'time averaged'. In areas where little sediment is normally preserved, such as in dry stream channels, collections of dry bone can accumulate and are then buried together following a catastrophic event (Behrensmeyer, 1975). In most cases in prehistory, however, a deposit covering several years is effectively an instantaneous event due to the vast amounts of time involved.

Another factor to take into account when interpreting fossil remains is taphonomic size bias since small bones and bones from small animals are regularly missing from fossil assemblages. Actualistic studies in East Africa have shown that in modern faunal communities bones from animals < 100 kg in weight experience much greater destruction than those from larger animals, and animals under 15 kg are particularly susceptible to exclusion from bone assemblages. Small animal bones can be more effectively chewed by carnivores, suffer from more breakage due to trampling, and are more vulnerable to weathering processes than bones from larger animals (Behrensmeyer et al., 1979). These pre-burial factors acting against the preservation of small bones are exacerbated by the fact that small animal bones may be further eliminated from a fossil assemblage by recovery bias. Small remains are more difficult to see than larger items and so could be passed over whilst surface collecting or excavating (Behrensmeyer and Dechant Boaz, 1980; Behrensmeyer et al., 1979). Small bones are also buried more easily and so may not be found on the same excavation surface as the larger bones within the same assemblage (Behrensmeyer and Dechant Boaz, 1980). It appears that the taphonomic processes operating on bones from small mammals could also play a part in removing small specimens, such as juveniles, of larger animals from fossil assemblages (Behrensmeyer et al., 1979). Therefore, without appreciating the circumstances under which small bones may disappear from an assemblage it would not be possible to interpret correctly the palaeoenvironment of the assemblage purely on the basis of faunal lists since whole categories of animals could be missing.

Despite the propensity of small bones to be omitted from a faunal assemblage, one of the most commonly recovered skeletal elements in the fossil record are isolated teeth, particularly molars (White, 1995). This is due to the fact that teeth are relatively dense and so do not suffer overly from trampling or physical weathering processes such as experienced in water transport. If exposed on the ground surface to heat extremes, however, they are likely to crack so making fossilisation doubtful (Behrensmeyer et al., 1979). Their small size, resulting in reasonably quick burial, is therefore an added advantage. This exemplifies the problem of understanding taphonomy, under certain circumstances and conditions different skeletal elements will survive and by recognising those conditions from the faunal assemblage it is possible to determine the original fossilising environment.

Pollen remains are also subject to alteration through many of the same taphonomic processes that also affect bone assemblages (Andrews, 1992). Whilst the outside wall, exine, of pollen grains is extremely resilient, it can be damaged by diagenetic changes as a result of bacterial action or through oxidation if the deposit in which it is preserved is subjected to drying out episodes (Williams et al., 1998). Sediments containing pollen from one time period can experience mixing of pollen from later or earlier periods through root or worm activity as well as through disturbance by humans (Williams et al., 1998). Therefore, sediments chosen for pollen analysis should be free from any obvious signs of degradation or mixing, although this is sometimes difficult to determine.

5.2 Methods of Analysis

Once an understanding of the taphonomic problems affecting a fossil assemblage has been attained, various methods of analysis are available to assist in the interpretation of the palaeoenvironment. The suitability of each method is dependent on the quality and quantity of the remains uncovered, as well as the specific detail required, and as such each technique suffers from its own particular shortcomings.

5.2.1 Taxonomic Analysis of Faunal Assemblages

One of the most obvious ways of reconstructing palaeoenvironments is to use faunal lists from excavated sites and surface collections. This technique could immediately run into problems, however, since it assumes both that the bones are associated and that the fossil remains have been correctly identified (Andrews, 1996). Whilst surface collections frequently involve picking up bones eroding out of a sedimentary layer, for instance on a cliff side, often their precise stratigraphic origin (i.e. three-dimensional co-ordinates) is not certain, hence shedding doubt on any environmental reconstruction. This is of particular concern for a worker with no first-hand knowledge of the site or the excavation techniques used.

Understanding of various taphonomic factors plays an important part in ensuring that environmental reconstructions based on faunal lists are accurate.

Juvenile bones and bones from small species often disappear altogether from an assemblage as a result of predator consumption and weathering, in addition juvenile bones have unfused epiphyses making them easier to destroy (Behrensmeyer and Dechant Boaz, 1980; Western, 1980). Also, not all species of animals die at the same rate and animals with high birth and death rates are more likely to be recovered from a fossil assemblage than more long-lived animals since more carcasses are produced and hence more are likely to survive the fossilisation process (White, 1995). Body size still plays a part, however, since elephants fall into the general category of low birth rates and associated longevity, but as a result of their massive size are still reasonably well represented in bone assemblages. Therefore, it is important to know what, if any, bones are likely to be missing from an assemblage.

When identifying fossils to taxa, there should not be too much uncertainty as far as well-preserved bone specimens are concerned. Unfortunately, under most circumstances the fossil remains uncovered are extremely poorly preserved with only mid-shaft fragments of long bones surviving and few diagnostic traits visible (Walker, 1980), although denser bones such as vertebrae and crania and teeth do tend to fare slightly better. The smaller the fragment of bone preserved, the more likely it is that it will be misidentified by workers who may know little about taxonomic groups outside their area of expertise (Walker, 1980). As a result, it is not unusual for fossil remains to only be identified to superfamily or even higher classificatory categories, although family level attributions are more common with some specimens, normally those with few close relatives, referred to generic levels. More accurate classifications, to species level, are only possible on extremely well preserved fossils whose genealogy is well known or for those individuals representing the only member of their genus. For instance, there is probably only one species of rhinoceros represented in the Upper Siwaliks of Pakistan and so all remains can fairly safely be referred to *R. sivalensis* (Dennell, pers. comm. 2000).

A further problem relating to identifying fossil specimens correctly involves the massive difficulty of assigning taxonomic groupings to specimens that are known only from the few parts of their anatomy, such as teeth or crania, which survive the fossilising process reasonably well. Associated with this is the problem of determining which of the isolated teeth, horn cones and post-cranial remains that have been recovered belong to the same species or genera. African suid lineages, for example, have been identified almost exclusively through morphological changes of third molars due to the lack of remains from other parts of the skeleton (White, 1995). Exactly when enough change has come about to refer specimens to new species, or even genera, is a matter of some debate. As a result, the genera and species identified are unlikely to represent true biological units (White, 1995) and, therefore, it may be safer to use higher taxonomic categories in identification where the likelihood of error is reduced.

Once a specimen has been correctly identified, more assumptions are brought into play. Modern-day relatives of the fossil, or similar animals if the genus has become extinct, are examined and the habitat type of the fossil is inferred from the preferences of the living animals (Andrews, 1996). This implies several things, firstly that the modern-day species is living in its primary choice of environment. To put this more specifically, that it has not been forced into a habitat not entirely suitable for it as a result of pressure from other animals, including humans. Alternatively, it may be living in another locality due to the absence of other animals, hence extending its range (Andrews, 1996). Secondly, and linked to this issue, that the type of vegetation occurring within the particular environment favoured by the modern relatives has not altered considerably between the time period of the fossil and the present. Also, that the habitat the fossil species occupied is comparable to environments found in the world today. Lastly, that there should be a correlation in environment between the two animals. That is this method does not take into account any flexibility in habitat type, or even complete ecological change, of the species (Andrews, 1996; Shipman and Harris, 1991). As a result, a variety of different environmental scenarios may be proposed for the same fossil specimens.

Reconstructing habitats for the Plio-Pleistocene in this way is probably more accurate than for previous periods since many fossil animals have close relatives alive today and so available for comparative analysis. This procedure is normally carried out for a considerable proportion of the fossil assemblage, since if various taxa all indicate the same or similar type of palaeoenvironment it makes it more likely that the reconstruction is accurate than if only one genus is used. There are a number of drawbacks to this method, including the problem of identifying specimens to family or generic levels, which can be addressed by using various other techniques to reconstruct palaeoenvironments.

5.2.2 Functional Analysis of Faunal Assemblages

It is possible to reconstruct the environment of a site by examining the skeletal morphology of part, or all, of the fossil faunal remains uncovered (Andrews, 1996). The fossil morphology is then compared with that of animals whose habitat preference is known; the reasoning being that animals with the same morphology, in terms of limb length or teeth structure, will inhabit the same environment (Andrews, 1996; Vrba, 1980). For instance, modern-day open-country bovids are characterised by cursorial features such as limb bones that are longer with respect to volume and have brain cases that are at an angle to the facial axis, and it is assumed that fossil bovids exhibiting these same characteristics would also have inhabited an open-country environment (Vrba, 1980). The environment implied by most of the remains is then postulated to be an accurate representation of the palaeoenvironment. This presupposes, however, that the function of anatomical parts has remained steady, whereas it is equally possible that the animals could have adapted to another environment, either completely or for only part of their day-to-day life, without any alteration in their skeletal morphology (Andrews, 1996). It also assumes that the fossil remains have been exceptionally well preserved in order to carry out this type of analysis, since it is extremely unusual to recover intact fossil limb bones and skulls.

Using tooth morphology to infer environments is a more complicated issue since many of the criteria rest on qualitative assessment such as differences in crown or cusp height, a factor that can vary with the age of the specimen due to wear and tear (Vrba, 1980). Also, as with bone morphology, dental traits may represent ancestral characteristics that are no longer functional. An example of this is found in the dentition of *Sivapithecus*, a late Miocene hominoid with thickly enamelled teeth. On the basis of the tooth morphology a diet of tough, brittle food was suggested, but microwear analysis showed that mostly soft foodstuffs were eaten (Teaford and Walker, 1984 cited in Andrews, 1996), indicating that a dietary change had occurred at some point in the evolution of *Sivapithecus* that had not been reflected in hard-part morphology (Andrews, 1996).

Examination of fossil teeth under the scanning electron microscope can also provide an indication of the type of environment in which the specimen lived. Analysis of wear scratches makes it possible to differentiate between forest and open-country herbivores, browsers, grazers and carnivores, including those who chew bone (Walker, 1980). However, whilst this investigation does provide highly sensitive diet information, it remains difficult to reconstruct environments from the data since suitable food resources could be found within numerous habitats (Andrews, 1996).

Functional, morphological analysis has been conducted on the sub-family Equini by Alberdi et al. (1995) who discovered a definite link between the size of an animal and the environment and substrate on which it lived, agreeing with Bergman's rule whereby body size varies clinally. Equini of small body size tend to be found in warm climates with biomass containing an important wooded component growing on hard substrate. On the other hand, Equini of large body size are found in cold climates where there is open biomass growing on a soft substrate (Alberdi et al., 1995). Therefore, by determining the body size of the fossil it is possible to gain an impression of the type of environment inhabited. Nevertheless, it can be extremely difficult to estimate the full size of an animal from fragmentary fossil remains, particularly if it is not known if the fossil had reached complete maturity.

Vrba has used functional analysis in order to reconstruct palaeoenvironments. She examined the relationship between antelopes (Bovidae) and the environment, choosing this family of mammals as a result of its taxonomic diversity, the fact that bovids frequently specialise in terms of the type of vegetational matter that they consume and the quantity of remains of this family that have been recovered in the archaeological record (Vrba, 1980). Believing that the morphology of the different bovid tribes would shed light on their preferred habitats (Vrba, 1980), she conducted correspondence analysis on antelope census data from various ecosystems within sub-Saharan Africa. This made it possible to determine from living assemblages which bovid tribes were preferentially associated with particular environments (Greenacre and Vrba, 1984; Vrba, 1992). She found that the wood to grass proportion within the environment was the primary factor affecting antelope habitat choice, with this trait apparently being heritable with the entire reduncine clade preferring open woodland habitats, and all alcelaphine and antilopine genera preferentially inhabiting open grasslands (Vrba, 1980; Vrba, 1992). *Oryx* and *Addax* were also normally found in open arid environments, such as grasslands, an unusual characteristic for the rest of their tribe; *Taurotragus, Raphicerus* and *Sylvicapra* were found in such a wide variety of habitats that their remains were not characteristic of either the amount of rainfall or habitat type (Vrba, 1980). Therefore, if a fossil bovid assemblage was dominated by alcelaphine and antilopine genera, it would appear likely that the palaeoenvironment would have been open grassland. Likewise, if large proportions of *Oryx* and *Addax* were recovered from an assemblage, an open, arid habitat would be indicated.

There are, however, certain circumstances in which the above relationships between proportions of bovid taxa and the environment may not hold true. The applicability of this technique is restricted to late Pliocene and Pleistocene assemblages when Bovidae experienced a marked increase in species types (Vrba, 1980). The remains being analysed should cover a reasonable amount of time, since catastrophic assemblages or those representing single season accumulations may give a misleading impression of the type of environment the species inhabited. The area from which the remains were recovered should represent only one climate zone, because otherwise the mixture of species will not give a clear indication of the habitat (Vrba, 1980). Lastly, many bovid tribes have restricted weight ranges and so the absence of certain genera from a fossil deposit does not necessarily mean that the environment was unsuitable, but rather may represent taphonomic size biases. Therefore, the type of taphonomic processes that the remains experienced should be understood since many processes operate on specific sizes or types of bones resulting in a biased assemblage. Further analysis of other fossils within the assemblage would indicate whether the deposit as a whole had suffered from size related biases (Vrba, 1980).

Shipman and Harris (1991) reassessed Vrba's original work and found that other bovid tribes were also representative of habitat type. Agreeing with Vrba's original conclusion that Alcelaphini were a distinguishing feature of open habitats (Vrba, 1980), they also felt that Antilopini fell within the same habitat grouping (Shipman and Harris, 1991). Tragelaphines (including kudu and eland) and aepycerotines (impala) showed a preference for bushy, closed habitats with reduncines (e.g. waterbuck, kob) and bovines (buffalo) generally found near standing water in wet, closed habitats (Shipman and Harris, 1991). Confirmation of the accuracy of these groupings was obtained by comparing the environment type suggested by the bovid analysis to that indicated by pollen analysis. Whilst there is a danger in reducing habitat types to such simplistic levels, it does at least present a starting point from which palaeoenvironments can be examined (Shipman and Harris, 1991).

5.2.3 Palaeoenvironmental Reconstruction Using Entire Mammalian Assemblages

Perhaps one of the safest ways of reconstructing palaeoenvironments is to utilise the entire mammalian fossil assemblage, as shown in Reed (1997, 1998), since the use of only one taxon may lead to ecological bias (Cooke, 1978 cited in Reed, 1998). Also, the more species used in a reconstruction, the more likely it is that the reconstruction will be accurate (Western, 1980).

The species richness (heterogeneity) and species abundance (equitability) of the assemblage are important factors to take into account since the dominance of the assemblage by a few species can provide an indication of the environment that they inhabited (Andrews, 1996). This can clearly be seen by examining modern fauna inhabiting harsh environments, such as regions that are extremely arid, cold or highly seasonal; the fauna tends to be dominated by a few species with low amounts of species heterogeneity (Andrews, 1996; Behrensmeyer and Dechant Boaz, 1980). Other environments also have distinctive animal communities; lowland or montane forests are characterised by a high taxonomic diversity in species of rodents, insectivores and primates, whilst floodplain, woodland/bushland and savanna communities comprise higher numbers of both ungulate and carnivore taxa (Boaz et al., 1992).

There are problems associated with examining the heterogeneity and equitability of an assemblage, however, as certain animals may be more likely to be preserved in a fossil assemblage than others and some animals are easier to recognise. Therefore, it is important to know the sort of fossil assemblage being examined so that estimates of the types of animals either missing or over-represented can be made (Andrews, 1996). Some species can be identified more easily from fragmentary remains leading to an apparent abundance in the assemblage that is not necessarily accurate (Behrensmeyer and Kidwell, 1985). Other species may suffer from high death rates again leading to over-representation in a fossil assemblage, while still others may inhabit areas, such as forests, where fossilisation rarely occurs and so would be under-represented in any analysis of the fossil remains (White, 1995). Areas where animals live during the hardest part of the year in terms of survival will provide many more carcasses than would normally be expected, particularly elderly, injured or juvenile remains. This is also true of areas surrounding waterholes. Since all animals require water at some point, carnivores frequently 'stake out' water sources to attack the animals when they come to drink (Behrensmeyer and Dechant Boaz, 1980; Western, 1980)

Reed (1996) utilised functional analysis of all macro-mammalian remains found within a fossil assemblage in order to determine the type of environment early hominids found in association with fossilised mammal remains inhabited (Reed, 1997). She found that by morphologically analysing the trophic and locomotor adaptations for all specimens within modern African macro-mammalian death assemblages it was possible to infer the ecological adaptations of each species and so reconstruct the environment (Reed, 1998). The mammal assemblage was divided into various groupings by using locomotor categories such as terrestrial, arboreal, aquatic, fossorial and terrestrial/arboreal. Further subdivisions were based on trophic variables to determine the kind of vegetation that the mammals consumed and hence the environment they were likely to inhabit; the categories were browsers, grazers, mixed feeders, frugivores, insectivores, meat eaters, meat-bone eaters, meat-nonvertebrate eaters and omnivores. The percentage of animals falling within each of the different categories would be a diagnostic representation of different habitat types. As quite basic trophic groupings were used, it would normally be possible to assign fossil remains to the correct group, hence making the technique suitable for palaeoenvironmental reconstructions (Reed, 1997). The premise underlying the technique was that using biological adaptations to infer palaeoenvironments would be more fruitful than employing taxonomic analysis since in distant regions of the same habitat different species, although normally the same tribe or sub-family, may occupy the same ecological niche (Reed, 1998).

After examining the variety of animals that occurred in several different modern-day habitats within Africa, she was able to ascertain the differing proportions of animals falling within the alternative locomotor and trophic categories (Reed, 1998). Several predictor patterns emerged; the percentage of animals using a terrestrial form of locomotion decreased as the amount of tree cover increased and vice versa, the opposite pattern can be seen in arboreal mammals. As would be expected, the percentage of aquatic mammals increased with the presence of water. The best trophic variable was the percentage of frugivorous mammals and when this was combined with the evidence for locomotor type it was possible to distinguish between differing ecological zones (Reed, 1997).

Her experiments showed that modern habitat types can be predicted on the basis of the functional anatomy of the mammals found within them, since trophic and locomotor variables can be assigned, and that this was also true of both life and death assemblages. This makes this technique applicable to the fossil record and so solves some of the problems associated with reconstructing environments on the basis of taxonomy, however, there are major constraints. The assemblage analysed must not have suffered greatly from the effects of taphonomy and should include representatives of the whole community (Reed, 1997). Fossil assemblages rarely show such a wide range of habitat preferences as indicated in this research and so the full spectrum of analysis may not be possible. The technique is still useful in determining what groups of animals are missing from an assemblage though, so indicating from what habitats the fossil assemblage could not have been derived.

5.2.4 Palynological Analysis

Palynology, or pollen analysis, is another important means of interpreting and understanding palaeoenvironments as it can provide a more direct way of reconstructing habitats than by inference from faunal remains. It is unusual, however, for both faunal and floral remains to be preserved in the same geological levels and so direct comparisons between the different mechanisms of reconstructing palaeoenvironments are not always possible (Andrews, 1996).

As with other environmental reconstruction techniques, palynology has its own drawbacks and pitfalls. Whilst climate does have a direct influence on vegetation patterns, vegetation types are also determined by soil characteristics, and these are slow to respond to any changes in climate. As well, competition from other plants affects the distribution of vegetation and so the absence of particular taxa does not necessarily indicate that the environment was unsuitable (Moore et al., 1991). As with faunal assemblages, taphonomic considerations play a part in interpreting any floral deposits, particularly the original location of the pollen and the mechanisms by which it arrived at the site, since these could all have altered the original species composition (Moore et al., 1991). Pollen is also produced on a seasonal basis and so multiple samples may be required to ensure that any analysis includes all pollen from within the sample area (Birks and Birks, 1980).

The most favourable situations for pollen preservation are those which are acidic and poorly aerated, such as peat bogs and lake or sea beds, where microbial activity is low, the pollen grains and spores can be quickly covered by a fine layer of sediment and there is little later disturbance (Birks and Birks, 1980; Moore et al., 1991). Pollen samples are normally obtained by coring through sediments in lakes etc or by extracting pollen and spores directly from exposed horizons, for example on archaeological excavations. Extreme care has to be taken to ensure that there is no cross-contamination with either other fossil deposits or modern pollen rain. The taking of multiple samples within a stratigraphy allows comparisons of the vegetation through time to be made, although how regularly these samples need to be taken is dependent on the level of accuracy required as well as the speed of deposition (Moore et al., 1991).

Only small amounts of sediment, about 1 cm^3, are required for analysis and this is sub-sampled from the original sample. The soil matrix surrounding the pollen grains and spores is then removed through various chemical processes allowing quicker identification and easier counting of the grains (Moore et al., 1991). Pollen grains and spores are then microscopically examined, under either light or scanning electron microscopes, in order to undertake identification. Variations in the exine, the outer coat of the grains, allows classification to either family or genera, only occasionally are species attributions possible, and this has profound implications for environmental reconstruction as species falling in the same genera may live in different conditions (Moore et al., 1991).

After the pollen from a sample has been identified, a tally needs to be undertaken. Routine counting involves totalling the number of grains or spores in each taxa falling along a transect until the required amount has been reached, it is not normal practice to count every grain within a sample. This quantity should be large enough to provide a representative sample of the pollen within the assemblage, including rare taxa, although this is frequently hard to judge as samples from different environments require differing sampling strategies (Moore et al., 1991; Williams et al., 1998). Since the pollen grains and spores counted only represent a sample of the original deposit, the amount of pollen present is usually expressed in relative proportions of either total pollen and spores or total arboreal pollen (Moore et al., 1991).

It is possible to estimate the total pollen concentration within a sample in several different ways. A known volume or weight of the main sample is prepared for analysis and from this a small sub-sample, again of known size, is examined under the microscope and all pollen grains and spores are then counted. The amount of each taxon present within the main sample can then be calculated on the basis of proportional analysis (Moore et al., 1991). Alternatively, a known amount of exotic pollen could be added to a sample. A sub-sample would then be taken and all exotic and fossil pollen present would be counted. The fossil pollen concentration would then equal: –

$$\frac{\text{Counted fossil pollen}}{\text{Counted exotic pollen}} \times \text{Exotic pollen concentration added}$$

(Birks and Birks, 1980)

The pollen data obtained from stratified deposits is normally represented graphically in a pollen diagram since it is the most effective way of presenting the complex information (Birks and Birks, 1980; Moore et al., 1991). The manner in which the diagram is constructed varies with each worker, although certain conventions are usually applied. The vertical axis represents the depth and stratigraphy of the sample, whilst on the horizontal axis the abundance of each pollen taxon is indicated (Moore et al., 1991). As mentioned previously, each taxon is shown as a percentage of the pollen sum, normally either total land-based pollen or total arboreal pollen. Aquatic plants are frequently omitted from the diagrams since they depict an environment not related to the rest of the pollen (Birks and Birks, 1980). In samples where there are taxon identified by both high and low values of pollen, the scale of the diagram can be exaggerated to allow the entire sample to be depicted on one graph. The proportions of each taxon can be illustrated by either a histogram at each sampling point within the stratigraphy with gaps in between or as a curve joining the points together. In many respects the histogram is more accurate since the amount of pollen present in the deposit falling between the sample points is unknown and the drawing of a curve suggests a level of certainty that is not justified (Moore et al., 1991). Multiple pollen diagrams are sometimes produced for each site as a way of ensuring that the various sampling procedures have left an accurate record. If all samples indicate the same general pattern then the researcher can be confident that any conclusions drawn are legitimate and are not artefacts of the analysis (Moore et al., 1991).

Unfortunately, different plants produce varying amounts of pollen and differ in how they disperse the grains and spores, wind pollinated plants produce considerably more pollen than those relying on insect pollination due to the inefficiency of the reproductive system (Moore et al., 1991). Therefore, a high proportion of grains from a certain taxon does not necessarily mean that it dominated the vegetation assemblage since a large amount of pollen could have been produced by a few plants. *Pinus* (pine) is frequently over-represented in pollen diagrams due to the vast amounts of pollen that are produced, the ease with which it is transported, and the fact that it preserves well in many situations (Heusser, 1988 cited in Combourieu-Nebout, 1995). In the tropics, the amount of arboreal pollen present in a sample frequently underestimates the density of trees, particularly in grassland or dry woodland areas, as trees produce much less pollen than grasses (Mworia et al., 1988 cited in Bonnefille, 1995). Therefore, it is necessary to understand the reproductive behaviour of various taxa in order to ensure that pollen spectra are correctly interpreted.

The presence of certain types of pollen within the sample is affected by the distance of the depositional environment from the plants. Pollen that is transported by wind is found over much greater distances than that transported by insects (Scott, 1984), and water-transported pollen can also be carried substantial distances if streams are present near to the plants (Williams et al., 1998). Therefore, any pollen sample will not only include pollen from the immediate area, but also pollen from around the sample area and deposited through water transport, pollen indicative of the regional vegetation and pollen transported from large distances (Jansson, 1966 cited in Williams et al., 1998). From this it can be seen that the percentages of pollen grains or spores illustrated in pollen diagrams do not correspond exactly to the vegetation they represent and as a result the diagrams need to be interpreted. Despite this problem, however, patterns of vegetation change at a location can be illustrated (Combourieu-Nebout and Vergnaud Grazzini, 1991).

Interpretations of palaeo-pollen diagrams are based on comparisons with modern spectra relying on the theory of uniformitarianism, whereby fossil plants are believed to have inhabited the same or similar kinds of environment as their modern counterparts (Birks and Birks, 1980; Bonnefille, 1985; Moore et al., 1991). Modern pollen samples have been collected from known environmental habitats and analysed in the same manner as those from fossil locations so that known standards could be developed (Moore et al., 1991; Williams et al., 1998). Plant genera associated in fossil assemblages are thought to indicate the same environment as a similar association found in modern-day environments (Behrensmeyer, 1985). The presence of certain pollen grains, termed indicator species, in assemblages provides an indication of the surrounding habitat when the pollen was deposited, based on the modern climatic preference of plant families. These taxa can normally only survive in very specific conditions, so suggesting a narrow ecological setting for the environment surrounding the pollen assemblage (Combourieu-Nebout and Vergnaud Grazzini, 1991; Moore et al., 1991). More generally, when Chenopodiaceae and Amaranthaceae occur in the same deposit it would normally signify aridity, and the presence of Gramineae generally points to openings in forest and woodland or wooded grasslands and grasslands (Bonnefille, 1995). Human disturbance of the natural environment, however, may make the use of comparative analysis unviable in some locations since the modern conditions would not be representative of any that occurred in the past (Birks and Birks, 1980).

Palynology can provide two levels of definition about past habitats. Initial interpretation of pollen diagrams provides information about the vegetation, that is the generic composition of the flora, and it is from this that inferences about the palaeoenvironment are made (Andrews, 1996; Moore et al., 1991). This inferential jump cannot always be made, however, since the vegetational structure may be ambiguous, especially if the pollen has accumulated from a wide area over a long time period (Andrews, 1996). Also, some plants are frequently found in more than one community type affecting the accuracy of any reconstruction (Birks and Birks, 1980). Therefore, broad habitat types, such as grassland or deciduous forest, may be the most accurate representation of palaeoenvironments that can be achieved.

5.3 Conclusions

The most overwhelming problem that needs to be overcome when reconstructing palaeoenvironments on the basis of floral and faunal data is the lack of modern analogues for certain environments, since in these cases comparative analysis cannot be used (Andrews, 1996). This is particularly true for assemblages excavated outside Africa, since most actualistic studies applicable to the Plio-Pleistocene period have been conducted in that continent. As a result, it is necessary to remain vigilant and not to impose modern analogues on unsuitable situations. Bearing in mind that most modern habitats have suffered from the influence of humans, particularly in Europe and Asia, the doubt always remains that an environment may not be reconstructed accurately (Reed, 1998). Also, in order to reconstruct palaeohabitats, modern-day habitats are often described in a simplistic manner, ignoring many of the more complex variables in order to be able to readily identify similar habitats in the past (Shipman and Harris, 1991). As a result many of the intricate interactions that occur in a habitat will go unrecognised in reconstructed palaeoenvironments.

Other problems of using mammalian evolution to interpret environmental change and to reconstruct the palaeoenvironment remain. The fossil record is not complete, and so a continuous record of climatic or evolutionary change is not possible (White, 1995). Some environments in which animals could have lived in the past may be entirely unsampled in terms of the fossil record, and so a particular aspect of a species habitat range may be unknown and so would not be included in a possible reconstruction. Wooded environments in particular are difficult to sample, in the case of modern census data this is due to the poor visibility whilst in the fossil record woodland is a poor environment for preservation (Western, 1980; White, 1995). Also, different species have different population densities, and so absence from a faunal sample may only mean that the species was rare and not that the habitat was unsuitable for the animal (White, 1995).

When assigning fossil remains to species, arbitrary taxonomic boundaries are created on the basis of apparent morphological change. This is as close as palaeontologists can get to identify species, but in reality the actual species break may occur at different places along the genus lineage. Therefore, reconstructing environments on the basis of taxonomic analysis may not be as accurate as using ecological or biological criteria (White, 1995).

Analysing assemblages in which both faunal and floral fossils are preserved undoubtedly remains the best way of attaining an accurate environmental reconstruction since the results obtained through faunal analysis can be compared with those reached through palynology. In this way, under ideal circumstances, pollen analysis could provide information on specific vegetation and climate, with faunal analysis presenting an indication of the preferred vegetation type of the animals analysed (Behrensmeyer, 1985). As a result of these problems, many environmental reconstructions represent the 'best fit' of modern data to the fossil record and as such are broad-brush reconstructions lacking much of the data known about contemporary situations.

Chapter 6

The Prism Dataset

As part of a long-term goal to understand global warming patterns, the Pliocene Research Interpretation and Synoptic Mapping (PRISM) project was developed within the U.S. Geological Survey. The objective was to reconstruct an interval of past global climate representing a period warmer than that of today's world. The period chosen was the Middle Pliocene, between 3.29 – 2.97 Ma, since it was the latest extensive period with warmer than modern climate but still geologically recent so allowing greater confidence in the methods used to reconstruct the climate (Barron et al., 1994; Barron et al., 1995; Dowsett et al., 1999). The period chosen was one in which there was a large amount of variability in climate, linked to the 41kyr obliquity cycle that dominates the Pliocene record (Tiedmann et al., 1994 cited in Dowsett et al., 1999). As a result, an estimate of the mean 'interglacial' conditions within the interval chosen was the data used, so minimising problems of site to site correlation across large distances (Dowsett et al., 1999). However, this does need to be taken into account when comparing more recent data with the PRISM databases since the same 'massaging' of data would not have taken place (Haywood et al., 2000).

The project reconstructions included sea levels and land-sea distributions, sea surface temperatures, sea ice and land ice distributions, land surface topography and vegetation distribution (Dowsett et al., 1999). Whilst several of these, particularly sea surface temperatures, sea ice and land ice distributions, were of no direct relevance to this research, the ability to use the PRISM datasets meant that a baselevel could be established to which later reconstructions of Pliocene and Early Pleistocene climate, and particularly vegetation, could be compared.

6.1 Pliocene Reconstructions

In order to reconstruct the Pliocene climate accurately, a variety of different data had to be obtained from fossil pollen, plant megafossil, palaeontological and palaeohydrological sources (Thompson and Fleming, 1996). This data then had to be related to a point on the earth's surface in a manner that allowed its integration with other sources. This was achieved by dividing the earth into a series of equal cells that could be identified sequentially. Due to computer storage requirements, as well as the general coarseness of the data itself, the size of the cells was restricted. The earth is already divided into 180° of latitude north and south of the equator and 360° of longitude east and west of the Prime Meridian and so by subdividing these units it was possible to achieve a suitable scale.

The initial reconstruction, PRISM1, was restricted to the Northern Hemisphere, due to insufficient data quantity from the southern regions, with information from 64 marine and 74 terrestrial sites. PRISM2 was a revision of the data portrayed in PRISM1, with additional coverage from the Mediterranean Sea and Indian Ocean to improve the resolution, and was also extended to incorporate the Southern Hemisphere. The increase in resolution meant that whilst the initial reconstruction was developed on a scale of 8° x 10° for each cell, the cell size in the revised reconstruction was only 2° x 2°, with each matrix of data being 90 rows long by 180 columns wide (Dowsett et al., 1994; Dowsett et al., 1999). The majority of the information was obtained from sites within the mid and high latitudes, predominantly within the North Atlantic and North Pacific Oceans.

6.1.1 Pliocene Coastline and Sea Level

The first step in reconstructing the Pliocene climate was to define the Pliocene coastline, since the reduction in ice volume at the two poles and associated eustatic changes would have resulted in a higher sea level than that of today (Barron et al., 1994; Dowsett et al., 1994). Whilst initial research suggested that a sea level rise of 35 m was appropriate for this period of the Pliocene (Thompson and Fleming, 1996), the revised reconstruction, PRISM2, based on the latest data, used a rise of 25 m (Dowsett et al., 1999).

Modern elevation data from the ETOPO5 five-minute topographic grid provided the foundation from which the Pliocene elevation could be calculated (Edwards, 1992 cited in Dowsett et al., 1999). The altitude of every ETOPO5 point in each 2° x 2° cell was examined to see if it fell at, above or below 25 m above modern sea level. If the majority of ETOPO5 points fell below the 25 m level the entire 2° x 2° cell was designated water, whilst if the majority of ETOPO5 points fell at or above 25 m the cell was classified as land (Dowsett et al., 1999). Whilst this was a crude way of discerning the Pliocene land-sea boundary, it did give general outlines of the continents since only low-lying areas, peninsulas and isthmuses were likely to have been affected by the change in sea level (Dowsett et al., 1994; Thompson and Fleming, 1996).

6.1.2 Pliocene Topography

Within the PRISM reconstructions, and especially within the area focussed on by this research, the topography was little altered from that of the modern day. This is principally connected to low levels of knowledge (Thompson and Fleming, 1996). In general, it was believed that Pliocene levels of elevation were not substantially different from those of today in most areas of the world. The major exceptions were in parts of the

Americas and Greenland, where most of the loss of height was a result of smaller ice sheets, and thus of no direct relevance. It was discovered, however, that the East African Rift Zone was likely to have been slightly higher in the Pliocene since the amount of rifting would not have been as great as that which has now occurred. Accordingly, the altitude in the Rift Zone area was increased by 500m from the modern situation (Dowsett et al., 1999). The topography for all other areas of the research area remained unaltered from that visible today.

6.1.3 Pliocene Vegetation Reconstruction

Once every cell had been categorised as either land or water, every land cell had then to be assigned a vegetation type. This was accomplished by simplifying a 1° x 1° modern vegetation grid. Twenty-two different types of vegetation were represented in this scheme, but this level of accuracy could not be expected for the Pliocene reconstruction (Matthews, 1985 cited in Thompson and Fleming, 1996). As a result, the vegetation was simplified into seven categories with all grassland counting as one category due to the difficulties in distinguishing the differences in the fossil record. The vegetation types were: desert, deciduous forest, evergreen forest, grassland (including steppe or savanna), rainforest and tundra (Barron et al., 1994; Dowsett et al., 1994; Thompson and Fleming, 1996). The initial 1° x 1° grid was then transformed to the 2° x 2° scale of the rest of the PRISM data by summing the four squares present in each 2° x 2° cell. The higher latitude pair in each cell was down-weighed due to the differences that occur in vegetation as a result of latitude. The modern vegetation type of each cell was then determined on the basis of which vegetation category dominated each cell, with only one vegetation type being assigned to each 2° x 2° cell (Thompson and Fleming, 1996).

The Pliocene vegetation reconstruction was achieved by examining the palaeoenvironmental evidence from the deep sea cores, sedimentary sequences and regional analyses locations and mapping the general vegetation trends. As with the modern vegetation map, only one vegetation type was assigned to each 2° x 2° cell (Thompson and Fleming, 1996). Due to the limited number of Pliocene sites available, the data obtained was not a representative sample of the world and should be treated as provisional. Furthermore, the age of many of the samples, particularly those obtained from terrestrial sites, was not known with any great confidence and adds a further area of uncertainty to the data since certain regions, such as Western and Southern Europe underwent considerable climatic change during the period in question (Barron et al., 1995; Dowsett et al., 1994; Thompson and Fleming, 1996). Both the PRISM1 and PRISM2 reconstructions were almost identical for the Northern Hemisphere and the PRISM2 reconstruction, with modern vegetation also pictured for comparison, is reproduced in Illustration 3. Since this research is solely interested in the Eurasian Northern Hemisphere, considerable portions of the earth are not portrayed.

As discussed above, in Section 5, interpreting palaeoenvironments and palaeovegetation is heavily reliant on the Principle of Uniformitarianism using primarily qualitative methods, and the PRISM reconstructions are no different (Dowsett et al., 1994; Dowsett et al., 1999; Thompson and Fleming, 1996). Bearing these points in mind, it is nevertheless clear from the vegetation maps in Illustration 3 that the Pliocene climate had to have been considerably different from that of today. The most obvious changes were the extension of coniferous forest over much of the modern tundra zone and, of particular importance to this research, the expansion of grasslands in the mid and low latitudes at the expense of the desert belt. This would appear to be characteristic of an environment where the temperature gradient between the equator and the poles was relatively gentle with a marked increase in temperature, compared to the modern situation, in the more northerly latitudes. Regions closer to the equator seemed to have enjoyed a moister climate allowing the expansion of grasslands, however, the extent of the deserts is difficult to ascertain since the data is almost solely based on inferences from nearby regions (Dowsett et al., 1994; Thompson and Fleming, 1996).

6.2 General Circulation Models

General circulation models (GCMs) can be used to assess the validity of the Pliocene climate and vegetation reconstructed by the PRISM project members. GCMs are used to simulate climatic conditions within certain prescribed boundary conditions. By using aspects of the PRISM datasets, such as sea levels, sea surface temperatures, vegetation etc, for the boundary conditions, an appropriate climate is generated by the model. This climate can then be compared to available geological and palynological evidence to assess its accuracy (Chandler, 1994; Chandler et al., 1994; Haywood et al., 2000; Sloan et al., 1996), since the results of the simulation are dependent on the boundary conditions (Sloan et al., 1996). However, discordance between the model results and the PRISM dataset may be due to problems integral to the GCM and not related to the PRISM data.

Results from the Goddard Institute for Space Studies (GISS) GCM and the National Center for Atmospheric Research (NCAR) GENESIS GCM suggested that whilst in some cases the PRISM data utilised was accurate, in other instances flaws were visible (Chandler, 1994; Sloan et al., 1996). The temperature estimates indicated by the PRISM project would appear to be correct, since the GCM returned warmer surface air temperatures at higher latitudes and cooler temperatures at the equator (Chandler, 1994; Chandler et al., 1994; Sloan et al., 1996). This was particularly true for the East African region, where temperature falls of 2 – 3° C compared to the modern situation were denoted by the GISS GCM. This cooling is in agreement with the palynological record from the area and confirms some of the PRISM conclusions (Chandler, 1994; Chandler et al., 1994). The Pliocene temperature structure also corroborates

Illustration 3. Changing Vegetation Types Within the Research Area. (Data from Dowsett et al., 1999).

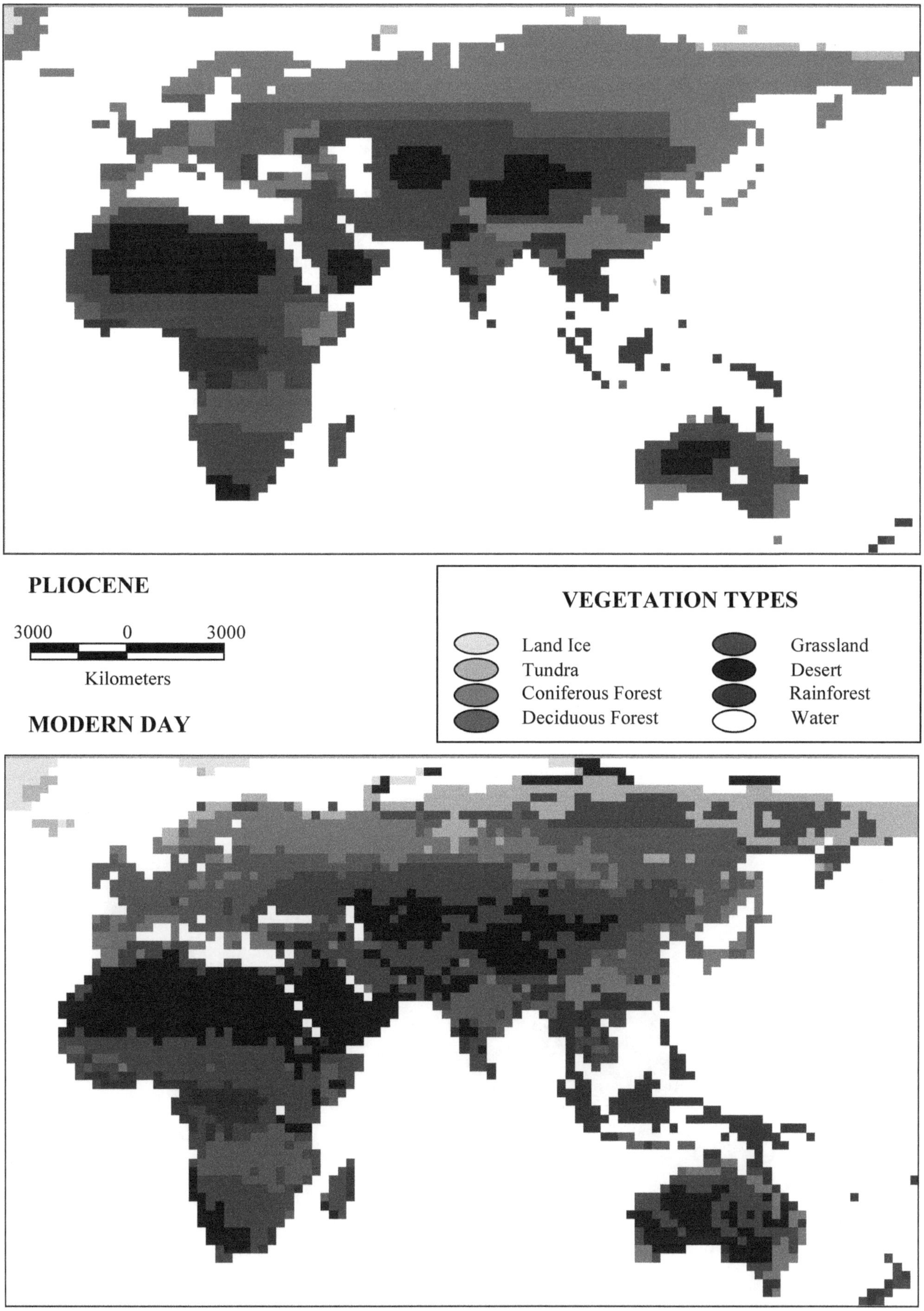

palaeobotanical evidence recovered from continental Europe where an increase in temperature over the winter is implied. This is likely to be due to changing albedo levels as a result of decreased winter snowfall (Chandler et al., 1994). However, the GISS GCM indicated summer drought conditions within the Northern Hemisphere, which, if true, would be seen in Pliocene terrestrial records. These records show no indication of drought; in fact summer seasons were frequently moister than in modern times. This discrepancy is most likely to be explained by problems associated with the model as hydrological cycles are difficult to replicate, however, it could also be due to inaccuracies in the PRISM reconstruction or difficulties in interpreting the fossil record (Chandler et al., 1994).

Some of the patterns seen in the GISS and NCAR GENESIS GCMs (Chandler, 1994; Chandler et al., 1994; Sloan et al., 1996) are repeated in the UK Meteorological Office (UKMO) GCM that used the PRISM2 data to prescribe the boundary conditions for the model (Haywood et al., 2000). Again, surface temperatures increased from modern ranges in mid to high latitudinal regions in both summer and winter, but decreased in regions of Africa and Southern Asia. Higher levels of precipitation were also recorded for equatorial and low latitude regions, particularly India, Arabia, Central and Western Africa. This was in accord with the PRISM datasets and conclusions, the NCAR GENESIS simulations, as well as various geological data (Haywood et al., 2000; Sloan et al., 1996). Once more, however, some inconsistencies between the UKMO and NCAR GENESIS GCM simulations and the PRISM dataset were apparent. These related to problems concerning oceanic upwelling off the North American coasts and the size and nature of the northern and southern polar ice sheets (Haywood et al., 2000; Sloan et al., 1996). Whilst these localities fall well outside the area investigated in this research, the GCMs have a global scale and inaccuracies in one region could affect results in other parts of the world. Therefore further work on these areas is needed in order to interpret the GCM results correctly (Haywood et al., 2000). Nevertheless, the fact that the UKMO GCM is in general accord with other GCMs, as well as the PRISM data, does imply a reasonable level of robustness in terms of the accuracy of the PRISM reconstructions.

6.3 Conclusions

The PRISM2 datasets have been constructed from a wealth of information covering many aspects of the fossil record and their use in several GCMs shows that they depict a reasonably accurate interpretation of the Middle Pliocene climate and vegetation. As a consequence, they represent an invaluable source of data from which hypotheses about Later Pliocene climate and vegetation and their impact on migration can be generated.

One of the ways in which hypotheses to be tested by this research were generated was by directly comparing the vegetational distribution of the Middle Pliocene and the present day, as is shown in Illustrations 3 and 4. This made it immediately apparent that certain areas, such as North Africa and Arabia, were more likely to be important regions for floral and faunal migrations than had been expected in the past, and as such represented areas where research should be concentrated.

Despite the advantages of the PRISM datasets, it was nevertheless important to verify aspects of the PRISM reconstruction, such as topography and vegetation, to ensure that assumptions made about the Middle Pliocene still held true in the Later Pliocene and Early Pleistocene. This substantiation is discussed in the following two sections.

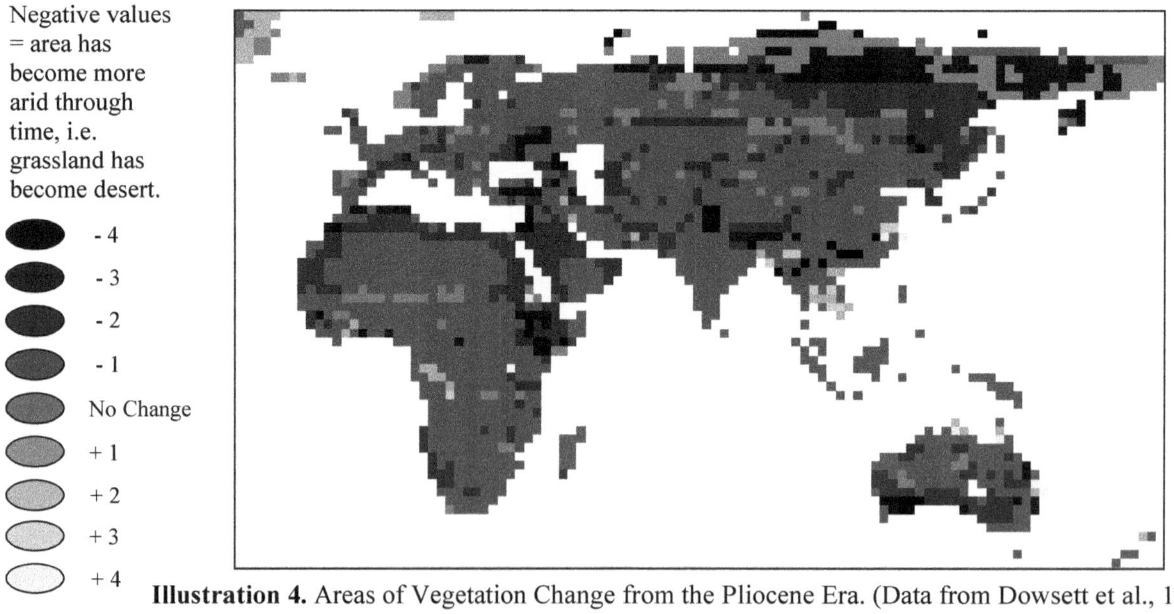

Illustration 4. Areas of Vegetation Change from the Pliocene Era. (Data from Dowsett et al., 1999).

Chapter 7

Late Pliocene and Early Pleistocene Geography and Associated Impact on Mammal Migration

The research area of North Africa, Europe and Asia underwent a period of great environmental and physical change within the timescale being considered. As a result, parts of the world looked dramatically different from how we see them today. A full understanding of Plio-Pleistocene geography is necessary to interpret how plants and animals, including hominids, were able to migrate effectively and colonise new areas. (See Illustration 5 for the location of sites mentioned within the text).

7.1 The Tibetan Plateau

The Tibetan Plateau has an area over $7x10^5$ km^2 (about the same size as Turkey), and is surrounded by several mountain chains including the Himalaya-Karakorum-Hindu Kush to the south and the Kunlun Mountains to the north (Dewey et al., 1988; Robinson, 1976). The extent to which the land barriers of the plateau and mountains would have disrupted animal migration either directly, or indirectly by affecting weather conditions, is much debated, with no clear consensus apparent.

The topographic phenomenon of an extremely high plateau bordered by even higher mountains was formed by the collision of the Indian and Asian crustal plates around 50 million years ago (Fort, 1996). Between 100 and 80 Ma most of modern-day Tibet was below sea level, as evidenced by the formation of Cretaceous limestone visible on the plateau (Molnar, 1997). As a result of plate tectonics the Indian plate moved inexorably closer, from a starting point off modern Madagascar, to the Eurasian landmass, leading to the collision and subduction of the Indian crustal plate underneath the Asian plate. The mantle, however, continued its northward movement and flowed under Southern Asia, due to the fact that it is denser, and composed of stronger rocks than the crust. This collision, which is still ongoing, led to forced upward movement resulting in mountain orogeny (Molnar, 1997).

The continental crust found underneath the Tibetan Plateau is considerably thicker than that found elsewhere on the planet and is responsible for supporting the great height of the plateau. The exact mechanism by which the thickened crust formed is under debate with two conflicting theories holding centre stage. Firstly, the crust to the north of the Indian plate could have been forced under Tibet along with the mantle. Secondly, the distance between Northern and Southern Eurasia could have

Illustration 5. Location of Important Sites Mentioned within the Text of Sections 7 and 8

Levantine Sites:
Bethlehem
'Ubeidiya
Latamne
Gesher Benot Ya'aqov
El – Ghab
Melekh Sdom
Jisr – Ash – Shaghour

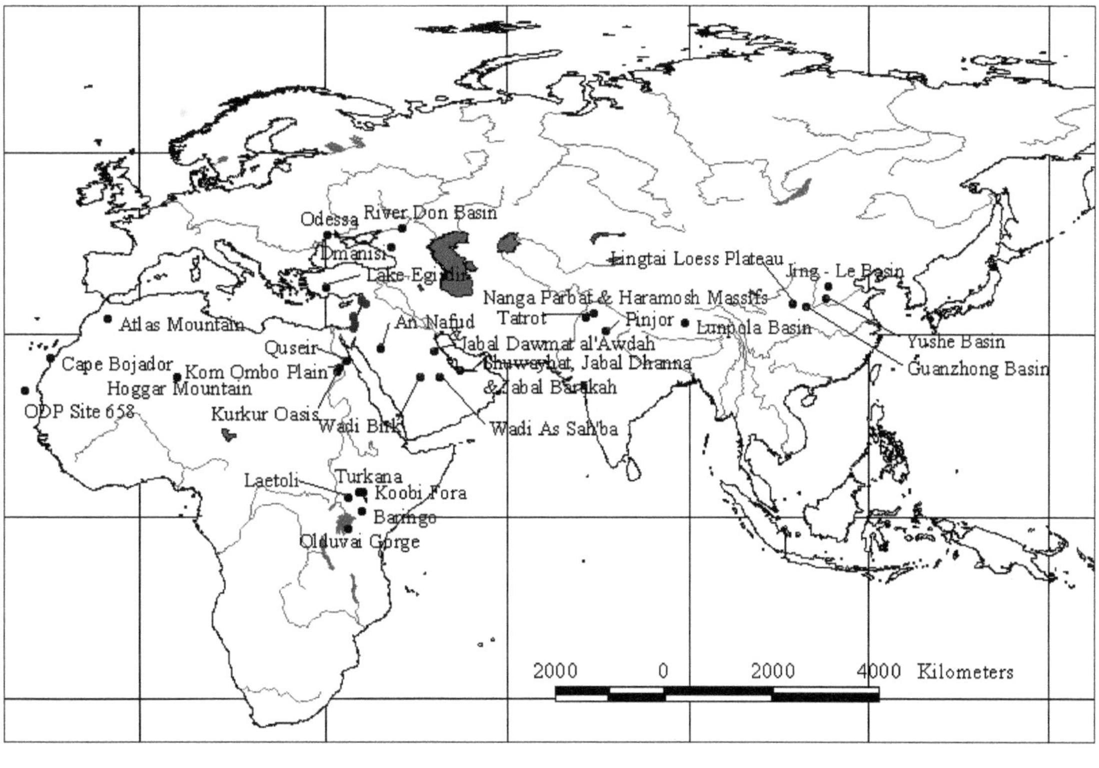

shrunk due to the forcing together of the two plates leading to folding of the crust beneath the plateau; palaeomagnetic research in Southern Tibet has indicated up to 1700 ± 600 km of crustal shortening between Siberia and Tibet (Molnar, 1997). Both these mechanisms cannot be entirely correct since it is impossible to account for the entire suggested crustal thickening by one of these methods alone, and it also seems unlikely that the whole of the Northern Indian plate could have been underthrust. It would therefore appear that whilst some crustal underthrusting could have occurred, the vast majority of the Indian crust would have been subducted and destroyed. The excess thickness of the crust could then have developed as a result of crustal shortening and the ongoing collision between the Indian and Asian plates (Molnar, 1997).

The Tibetan Plateau is currently at an average altitude of 4500m, with some individual peaks at a much higher elevation (Fort, 1996). The timing of this uplift is another contentious issue. There are currently two schools of thought, those who believe the plateau is a result of relatively recent uplift and those who think that the high elevation was attained in the Miocene (Fort, 1996).

The principally Chinese supporters of the 'recent uplift theory' cite evidence based on the floral and faunal remains found in Tibet, as espoused by Hsü in 1976 (Hsü, 1976 cited in Fort, 1996). They believe that in the Pliocene the Tibet-Himalaya complex lay at a low altitude of around 1000m and rose to 2000m by the beginning of the Pleistocene. The rate of uplift then increased so that the plateau reached its current height of 4500 – 5000m by the end of the Pleistocene (Fort, 1996).

Lower Miocene flora from the Lunpola Basin indicates that differences in altitude already existed with parts of the plateau exceeding 2000m and valleys at only 1000m (Xu Ren, 1981 cited in Shackleton and Chang, 1988). During the Miocene-Pliocene period, there is palynological evidence to imply that much of the plateau was dominated by tropical and subtropical climates, with woody steppe vegetation covering the peneplain. A typical *Hipparion* fauna has also been recovered from the Jilong Basin providing what is thought to be (but see below) further evidence of the prevailing low altitude (Chi-yuen Wang et al., 1982; Fort, 1996). Differential uplift was already occurring, however with the southern slopes of the Gangdise Range at Moincer now at more than 1500m and covered with deciduous, broad-leafed forests (Guo, 1981 cited in Shackleton and Chang, 1988; Xu Ren, 1981 cited in Shackleton and Chang, 1988).

The Late Pliocene period is synonymous with global climatic cooling, yet tropical and subtropical fauna and floral remains have been recovered from the plateau area, leading to suggestions that it was still low to moderate in altitude, despite the fact that the first glaciers would have developed in the Himalayan region at this time. This would further confirm that various areas of the Tibet-Himalaya complex were uplifted at differing rates (Fort, 1996). Rapid uplift is thought to have started in the Middle Pleistocene due to the widespread Nieniexongla glaciation, and the association of *Pinus* and *Quercus* together with *Alnus*, *Tsuga*, *Abies* and *Betula* during the interglacials is understood to confirm an altitude of about 3000m (Hsü, 1976 cited in Fort, 1996).

Advocates of the rapid uplift theory maintain that the final phase of uplift in the Upper Quaternary was the most rapid with an average increase of around 1500m in 100,000 years and led to extensive glacier development (Fort, 1996). This model proposed by the Chinese researchers therefore implies extremely high net rates of uplift in the last two million years, with an average uplift rate of 1 mm yr^{-1} (Chi-yuen Wang et al., 1982). This uplift could have occurred through a combination of tectonic processes as well as isostasy, whereby the erosion of deposits from the plateau surface would have led to the surface rebounding as part of the pressure forcing the plateau down was released (Molnar and England, 1990).

Several recent investigations of the Tibet-Himalaya complex throw doubt on the traditional Chinese interpretation of uplift timings. Much of the evidence previously quoted is open to more than one interpretation and newly noticed phenomena, such as east-west extension, have failed to be explained (Blisniuk et al., 2001; Coleman and Hodges, 1995). The main problem would appear to be that several of the supporters of the 'recent uplift theory' have failed to take into account the major climatic changes that have occurred since the Late Miocene. Many features that have been interpreted as being a result of uplift could actually have been caused by climate change (Fort, 1996; Molnar and England, 1990), e.g. Jian Wang et al., 1999.

Essentially, the 'rapid uplift' hypothesis to explain the timing of the Tibetan unroofing relies heavily on interpretation of floral remains to determine the height of the plateau. In order to be able to use palaeobotany to infer ancient altitude, the climate suggested by the plant remains needs to be resolved. If the region was originally at or near sea level, then the initial climatic state needs to be determined and the difference between the two regimes then related to altitude (Molnar and England, 1990). If the recent cooling of the earth is not taken into account, the suggested altitude could be out by as much as 1500m, since plants that today only grow at low altitudes could previously have survived at much higher elevations (Molnar and England, 1990), e.g. Jian Wang et al., 1999. This complicated, qualitative process of determining ancient altitude is open to many different interpretations and so whilst palynology remains a useful way of indicating the general climate, the drawbacks of the method need to be clearly understood (Harrison et al., 1992).

In geomorphology it is a commonly recognised phenomenon that uplift of the ground with respect to the base level of a water course gives the stream or river greater erosional power leading to the development of deeper valleys (Molnar and England, 1990). This same effect can be achieved through glaciation, however, and during the Plio-Pleistocene period the Tibet-Himalaya group experienced extreme and probably repeated glaciation. It would seem, therefore, that the deep incisions of the plateau and mountain surfaces giving the area a 'juvenile' appearance were a result of glaciation rather than uplift since the erosional activity of the glaciers would have removed any evidence of the earlier land surface (Molnar and England, 1990).

Other factors leading to scepticism about the 'recent uplift theory' include the following. The palaeoecology of *Hipparion* is not well established and so to determine maximum altitude on the basis of this fossil species would appear to be somewhat naïve as the exact species requirements are unknown (Fort, 1996). Also, the 1985 Geotransverse of the plateau from Lhasa to Golmud and field excursions from Lhasa to Kathmandu in 1981 and 1986 by Shackleton et al. (Shackleton and Chang, 1988) failed to find any evidence of a Pliocene peneplain across the Tibetan plateau, so disproving the theory of a relatively level plateau at low elevation at the start of the Pliocene, which was then later elevated (Shackleton and Chang, 1988).

The above data would appear to show that it is extremely unlikely that the Tibetan Plateau and surrounding mountain chains achieved their great height through recent rapid uplift, which would suggest that the Miocene uplift hypothesis is more accurate. As mentioned previously, it is the extremely thick continental crust present below the plateau that has allowed the high elevation to be achieved. This is an inherently unstable system, however, and once lateral gradients have reached sufficient strength within the earth's mantle the crust will begin to sink, causing the height of the plateau to fall. Evidence from remote sensing and Landsat images would appear to show that this stage has already been reached (Molnar, 1997).

Despite the inability of the system to support any more growth in height of the plateau, the crustal shortening that initially built the plateau is still continuing as a result of the sustained movement of the Indian plate northwards. The energy created by this movement has to be utilised and is seen in the development of normal faults running north-south along the plateau, cutting across all earlier structures, and causing an extension of the plateau in an east-west direction (Fort, 1996; Molnar, 1997). In order for this extension of the plateau to take place, however, the earth's crust needs to be relatively thin, a condition that could not have been met without some decrease in altitude having been achieved (Fort, 1996). The obvious assumption to be drawn from this is that the plateau is no longer at its maximum height (Molnar, 1997). By dating the start of the east-west extension of the plateau through the normal faults, it should therefore be possible to assign a date to when the plateau was at its maximum elevation (Blisniuk et al., 2001; Searle, 1995).

Most estimates of the start of east-west extension of the plateau have been based on chronostratigraphic data. For instance, Late Miocene basin sedimentation is shown to overlie north-striking faults in the Thakkhola graben suggesting that east-west extension had commenced by the Late Miocene (Fort et al., 1982 cited in Coleman and Hodges, 1995), whilst dates of 11 – 5 Ma have been suggested for the Yangbajian graben on the western flank of the plateau (Copeland and Harrison, 1993 cited in Coleman and Hodges, 1995; Pan and Kidd, 1992 cited in Coleman and Hodges, 1995). More recent ^{40}Ar/ sup ^{39}Ar studies of faults within the Annapurna range near the Thakkhola graben have provided a minimum age for east-west extension of approximately 14 Ma at the southern edge of the plateau (Coleman and Hodges, 1995). Before it can be demonstrated conclusively that the entire plateau had commenced east-west extension by the Middle Miocene, it will be necessary to date accurately the other fault systems in the area. It does seem certain, however, that the Tibetan Plateau had reached its maximum elevation some considerable time before the global climatic changes of the Plio-Pleistocene (Blisniuk et al., 2001; Coleman and Hodges, 1995; Searle, 1995).

It seems likely, though, that the Tibetan Plateau and the surrounding mountains did not move as one cohesive unit, but rather suffered from differential uplift (Fort, 1996). Sediments to the north of the Greater Himalaya have been found to dip northwards at anything from 5° – 15°, with the tilting eventually flattening out some 30 km north of the range. This would indicate an uplift of something in the order of 2 km since the sediments were originally laid down in the Mio-Pliocene. This uplift was, and continues to be, achieved by longitudinal basins forming to the north of the Greater Himalaya due to the ongoing thrusting of the Indian plate (Fort, 1996).

Confirmation of the differing uplift rates between the Tibetan Plateau and Himalayas can be seen in the river network, where all major rivers run to the north of and parallel to the Greater Himalaya whilst many antecedent rivers flow southwards across the main ridges of the mountains. This would appear to suggest that the rivers originally flowed downslope from the Tibet Plateau to the Ganges Plain before the rising Himalayas blocked their course (Chi-yuen Wang et al., 1982; Fort, 1996). The Indus River is one of the main antecedent rivers found in this region whose original drainage course of the Late Eocene was disrupted by the uplift of the Himalayas (Searle and Owen, 1999). Its source is north of Mount Kailas (very close to the source of the Ganges and Brahmaputra) from where the river cuts through all the ranges in North-West India and North Pakistan before flowing into the Arabian Sea. The high rates of uplift in this region have provided the river with enough erosional power to incise its paths deeply into the mountain. This can be seen between the Nanga Parbat and Haramosh massifs (35° 30' N, 74° 30' E) where the river has carved a steep gorge and the steepest gradient of the Indus is to be found together with the largest single slopes on earth (Searle and Owen, 1999).

It is also thought that the additional height of the Himalayas over Tibet could be a result of intense erosion of the Himalayan peaks leading to isostatic readjustment (Chi-yuen Wang et al., 1982). That is, once the heavy weight is removed from the tops of the mountains, the mountains become buoyed up due to isostatic rebound (Williams et al., 1998). Fission-track dating has shown that several areas of Tibet have suffered from large erosion rates during the Pleistocene and Holocene with about 10 km of rock removed from Nanga Parbat since 10 Ma and more than 6 km of this occurring in the last 1.3 Ma (Zeitler, 1985 cited in Searle and Owen, 1999). It would seem that uplift of 0.7 – 1.5 mm yr^{-1} for the Himalayas would appear probable based on evidence from fission-track analysis of minerals. This leads to an annual difference in uplift with the Tibetan Plateau of 0.2 mm yr^{-1}, further verifying that at one stage, about 2 Ma, the Himalaya belt was actually lower than the plateau (Chi-yuen Wang et al., 1982).

The evidence presented shows that for the entire timescale covered by this research the Tibetan Plateau would have played an important part in the geography of East Asia. Also, the Himalayan belt would have become increasingly more significant as it continued to be uplifted. Not only do the plateau and mountains represent physical barriers that would have to be crossed or circumvented by colonising plants and animals, they also had a massive impact on the monsoon climate of East Asia from the Miocene period onwards.

In terms of the model simulation, it was decided to maintain the modern-day height of the plateau in the Pliocene reconstruction, as was also done in the PRISM2 reconstruction (Dowsett et al., 1999), since any variation in altitude between the time periods would have been negligible. The height of the Himalayas has been halved from that of the present day in accordance with the view that the Himalayas and Tibetan Plateau were uplifted at different rates with the mountains being lower than the plateau during the Pliocene. This is at variance with the reconstruction proposed by the PRISM2 project, where modern-day values have been utilised throughout this area (Dowsett et al., 1999), but is supported by the evidence proposed by Wang et al. (1982). As seen shortly, this would have had major consequences on the region's monsoonal system.

7.2 East Asian Monsoon Circulation

The term monsoon refers to the seasonal reversal of winds, either from the Indian Ocean or continental interior, and so does not only apply to the rainy season (Webster, 1987). At the present time, during the summer, wind is blown inland from the cool ocean due to the differential pressure systems that exist over the continent and ocean. The associated rainfall is torrential since the warm, moist air that collected over the oceans is uplifted by convection currents as soon as land is reached, resulting in severe thunderstorms (Cohen, 1972). In the winter the wind direction is reversed as wind flows out of the high-pressure system over Siberia and into the low-pressure systems over the Indian and Pacific Oceans. This is a relatively dry period with cold air flooding south (Cohen, 1972).

The Tibetan Plateau adds a more complex direction to this scenario. The plateau is extremely high at 4000 m average altitude, with some peaks reaching more than double this height (Murakami, 1987). As a result, it represents a heat source that contrasts with the surrounding atmosphere forming a small high-pressure system during the summer months. This is an important influence in the 'wet' monsoon affecting India and parts of Pakistan, since the difference in pressure draws moist air into the continent from the oceans. During the winter the plateau acts as a barrier to westerly air currents, with the flow of air separating into northerly and southerly branches at low levels (Murakami, 1987). This has several effects in that it adds extra force to the downdraft of the Hadley Circulation, enhances surges of cold air from Siberia carrying dust from the northerly Chinese deserts to the Chinese Loess Plateau, intensifies the westerly jet over East Asia and leads to greater rainfall in equatorial areas of the South China and Indonesian Seas (Murakami, 1987; Tungsheng Liu et al., 1999). However, determining exactly when the modern-day pattern of wind circulation developed is controversial.

There would appear to be no question that the increased height of the Tibetan Plateau and Himalayas is intimately related to the commencement of the monsoon weather system in East Asia, due to the effect the Tibet-Himalaya complex would have had on global wind circulation. The height of the plateau would have disrupted west to east airflow, so initiating the monsoon wind patterns (Jablonski, 1993; Jablonski et al., 2000; Searle and Owen, 1999). Judging from recent evidence, however, the maximum height of the plateau was reached c. 14 Ma, much earlier than the development of the monsoon, which was thought to have commenced at 8 Ma (Molnar et al., 1993). However, it is possible that the monsoon did develop earlier, before experiencing a marked strengthening at 8 Ma (Quade et al., 1989). The height of the plateau is important in generating a heat source at high altitude, a major factor in the East-Asian monsoon system, and the threshold elevation needed to achieve this would have been reached in the Late Miocene. Today, the Himalayas represent the northernmost barrier to the monsoon with the arid Tibetan Plateau lying in the rainshadow to the north (Searle, 1995; Searle and Owen, 1999).

It is thought that major loess deposition commenced at about 2.5 Ma in North-Central China as a result of the East Asian winter monsoon wind picking up sediments from the arid interior of north China and Siberia, suggesting a strengthening of the winter monsoon at that time (Jule Xiao and Zhisheng An, 1999; Murakami, 1987). However, recent grain size analysis of samples of red clay dated to 7.05 Ma from the Lingtai loess plateau (35° 00' N, 107° 30' E) and found underlying the loess, showed similar sedimentological characteristics to loess deposits indicating that this too was a wind-blown deposit (Ding et al., 1999; Ding et al., 2000). Whilst many palaeoenvironmental analyses remain to be conducted on the samples collected from Lingtai, some indications of monsoon evolution can be suggested from observations of the pedogenic features found within the loess (Ding et al., 1999).

Sediments dating to 7.05 – 6.2 Ma are red/brown in colour with a weak, sub-angular blocky structure and few clay skins suggesting that the summer monsoon was still relatively weak. By 5.5 Ma clay skins were more common and the texture of the clay was moderately sub-angular, showing that the monsoon had started to strengthen. The period from 5.5 – 3.85 Ma had a strong summer monsoon as evidenced by many thick clay skins within the clay loess. This was followed by a short period of weakening in the summer monsoon, although by the end of red clay deposition at 2.6 Ma the strength of the monsoon had increased again (Ding et al., 1999 cited in Ding and Yang, 2000). Whilst these were preliminary interpretations of palaeomonsoonal climate, they do show that the monsoon did not increase or decrease in strength in a linear fashion (Ding et al., 1999).

One major difference that has been determined between the red clay and the overlying loess was the direction of the wind depositing the material. The grains within the red clay do not show a variance in size from north to south across the plateau area, whilst the sediments forming the loess show a marked difference in grain size with the grains to the south being considerably smaller than those to the north (Ding et al., 1988 cited in Ding et al., 1999; Liu, 1985 cited in Ding et al., 1999; Ding et al., 2000). As it is known that loess is transported by north-westerly monsoon winds, it has been inferred that the red clay was deposited by westerlies (Ding et al., 1988 cited in Ding et al., 1999). Therefore, the shift in deposition from red clay to loess could represent a change in wind circulation and hence monsoonal strength at 2.6 Ma. There is no doubt, however, that the initiation of the monsoon system occurred much earlier; further confirming that the Tibetan Plateau had already reached its threshold altitude (Ding et al., 1999), although what this height may be is unclear.

Knowledge of monsoonal circulation is important in understanding the complexities of East Asian climate and preservation of sediments since this is the dominant weather system over a large proportion of the research area, see Illustration 2. It is clear that for the timescale being considered the summer monsoon was active, although its strength would appear to vary. The lowered height of the Himalayas at the start of the Pliocene probably allowed summer rainfall to reach further inland on the Tibetan Plateau, perhaps resulting in a change of vegetation from what we see growing today as the level of precipitation would decrease with the ongoing mountain orogeny. The discovery of loess deposition prior to 2.5 Ma indicates that the winter monsoon is also an ancient feature of the climate in this area, although the changes in accumulation patterns do show that the system has not remained stable indefinitely. Therefore, any colonising plants or animals in the Plio-Pleistocene period would have faced an environment not dissimilar to that experienced in the same area today, with torrential rain falling for one part of the year leading to massive floods and huge movements of sediment, followed by an extremely dry period where drought and even famine were strong possibilities (Webster, 1987).

7.3 Grassland Expansion

Another environmental factor that became apparent in the Late Miocene and has also been attributed partly to the change in altitude of the Tibet-Himalayan complex, is the development and expansion of C_4 grasslands (Pagani et al., 1999). Plants fall into three different groups depending on their carbon isotopes, of which only two, C_3 and C_4, are relevant to the discussion since CAM plants comprise a minority of the world's vegetation. C_3 plants include almost all trees, shrubs, herbs and grasses growing in cool climates, at high altitudes or in wet conditions. C_4 vegetation includes some shrubs, but is mostly represented by grasses growing in tropical, warm climates with wet summer growing seasons, in open environments and occasionally in forests where there is a break in the canopy (Cerling, 1992; Quade et al., 1995; Quade et al., 1989).

Stable carbon isotopic analysis of mammal tooth enamel, soil carbonate and organic matter provides a mechanism through which it is possible to identifying whether C_3 or C_4 plants dominated an assemblage (Cerling, 1998; Quade et al., 1995). The $\delta^{13}C$ levels of C_3 and C_4 plants are very different with the average $\delta^{13}C$ values being -25‰ and -11‰ respectively. However, when isotopically analysing mammalian teeth, enrichment of the teeth by ^{13}C has to be accounted for, and so to unambiguously identify C_4 vegetation via this method a $\delta^{13}C$ level of -8‰ is required (Cerling, 1998). Isotopic fractionation also plays a part in altering the $\delta^{13}C$ levels of soil carbonates. Soils that developed under C_3 vegetation have $\delta^{13}C$ values in the range -10‰ to -12‰, whilst those formed under C_4 grasses have $\delta^{13}C$ values from -3‰ to +1‰ (Cerling, 1992).

Examination of mammalian tooth enamel from remains older than 8 Ma across varying continents has shown that all animals were consuming a diet comprised mostly, or entirely, of C_3 vegetation. By 6 Ma mammals from environments of less than 37° latitude across the major continents, but excluding Europe, were found to have a diet more dominated by C_4 vegetation (Cerling, 1997). This does not necessarily mean that only C_4 vegetation was available, however, since animals strengthen isotopic values by selective feeding due to the fact that not all animals are capable of feeding on a mixture of C_3 and C_4 vegetation as the leaf protein is 'packaged' in different ways (Cerling, 1997; Cerling, 1998).

Paleosols from several of the hominid fossil localities in East Africa have also been isotopically analysed in order to determine at what stage grassland expansion occurred. Before 1.7 Ma none of the localities of Laetoli, Olduvai Gorge, Koobi Fora, the Turkana Basin or the Baringo Basin had an environment that included more than 50% C_4 vegetation (Cerling, 1992). Whilst the environment would have to have been open to a certain extent in order for C_4 plants to survive, the landscape would have looked extremely different to the C_4 dominated grasslands present there today (Cerling, 1992). The most notable area for C_4 expansion is northern Pakistan, where before 7.4 Ma the carbon isotopic values recovered from pedogenic carbonates suggest a vegetation of closed canopy forest dominated by C_3 vegetation. By 5.0 Ma this had changed dramatically with C_4 grasses comprising the vast majority of the vegetation (Quade et al., 1989). The change here, as well as worldwide, has been interpreted as being a result of a modification in climate, specifically a reduction in the level of CO_2 within the atmosphere and the onset of seasonal monsoonal rainfall, leading either to invasion of the area by C_4 grasses or, alternatively, the evolution of this group of plants (Quade et al., 1989).

C_4 plants are much more tolerant of low CO_2 levels than C_3 plants and so the widespread expansion of this group of plants has been explained by a reduction in CO_2 levels, as mentioned above (Cerling, 1998). The extensive weathering that occurred in the Tibet-Himalayan region as a result of its uplift is thought to have lowered atmospheric CO_2 levels sufficiently to favour the expansion of C_4 grasses (Ding and Yang, 2000). The attribution of the expansion of C_4 grasses to changes in CO_2 concentration is contentious, however, since evidence for changing levels of CO_2 is difficult to obtain (Pagani et al., 1999). Recent estimates have suggested that CO_2 pressures stabilised at pre-industrial levels at around 9.0 Ma, with no apparent change at the time of C_4 expansion. Therefore, it would seem that other factors were also responsible for the expansion of these plants. Today C_4 grasses thrive in areas of aridity and seasonal precipitation and so it would appear likely that low-level aridity was at least partially responsible for their expansion (Pagani et al., 1999).

This theory is reinforced by the fact that the expansion of C_4 grasses occurred at discrete intervals across East Asia with the Pakistani grasslands evolving around 7 Ma whilst in Tibet, on the Chinese loess plateau, the same expansion did not occur until 4 Ma. If the expansion of the C_4 grasses was to be explained by changing levels of CO_2 worldwide, it is thought that expansion should occur synchronously at the same latitude (Ding and Yang, 2000). The extreme altitude of the Tibet plateau may help explain why C_4 plants were unable to colonise the area until relatively late in their history, as the cool temperatures of the plateau would have favoured the dominance of C_3 plants over C_4 grasses (Ding and Yang, 2000) although, when CO_2 levels are extremely low C_4 photosynthesis is favoured, even at low temperatures (Cerling, 1998).

It would seem likely, therefore, that the intensification of the monsoon weather system at 7 Ma encouraged the development of C_4 grasses in Pakistan as a result of the seasonal monsoon rainfall (Hoorn et al., 2000). The expansion in Tibet was likely to have been delayed due to the low temperatures (Ding and Yang, 2000; Quade et al., 1989) and it has been postulated that it was necessary for CO_2 levels to fall further before C_4 grasses were able to expand into the area (Cerling, 1997), although other factors could also have played a part. The delay in expansion over the plateau is another factor confirming that the plateau had to have reached a high elevation prior to 7 Ma, since it is almost impossible to postulate another reason why C_4 plants should have colonised the area so late.

By the Plio-Pleistocene C_4 grasses were widespread over large portions of Asia and were also forming an important component of the East African vegetation. This is in marked contrast to the situation visible today, where many of the former grasslands have become extremely arid and desert-like. This difference is visible on the vegetation reconstruction proposed for the Pliocene by the PRISM2 project and shown in Illustrations 2 and 3 (Dowsett et al., 1999), and further confirmed by floral and faunal analysis from sites across the region.

7.4 The Nile River

The River Nile is the only permanent watercourse within the modern Sahara Desert and is the longest river in the world, measuring 6,670 km from Rwanda to the Mediterranean. It is composed of three different river systems, the White Nile, Blue Nile and Saharan Nile, with the Atbara-Khor Gash system comprising a separate, but linked, unit, crossing vegetation as diverse as tropical rainforest, savanna and desert (Butzer et al., 1968). The familiar Nile drainage system, with its typical summer flooding, was not established until 25,000 years ago. A proto-Nile draining into the Mediterranean Sea had been in existence since the Early Miocene, but the White Nile and Blue Nile did not merge with the Saharan Nile before the Early Pleistocene (Butzer et al., 1968).

Up until the Late Pleistocene the part of the Nile south of the Sudan-Ugandan border drained westwards towards the Congo Basin and most of Sudan and Western Ethiopia had internal drainage in the Paleo-Sudd basin until the Middle Pleistocene (Butzer et al., 1968). The drainage pattern of Northern Ethiopia and Eastern Sudan was to the east, flowing towards the Red Sea, until the Middle Pleistocene by which time the tectonic situation in the rift valleys had altered to such an extent that that outlet was no longer open. Until the Early Pleistocene, the catchment area for the proto-Nile was towards the east near Aswan, rather than from the more southerly Nile rivers (Butzer et al., 1968).

Therefore, for much of the Plio-Pleistocene the river would have represented less of a barrier to faunal dispersals from east to west, and also less of a corridor from north to south; a factor that models of human migrations between Africa and Asia need to take into account.

7.5 The Red Sea

During the Oligocene and Early Miocene the Tethys Sea covered a large proportion of modern-day Europe and linked the Mediterranean and Indian Ocean east of Arabia. Doming of the Western Arabian Shield had commenced, leading to rifting at the centre of the shield and so establishing the Red Sea basin, but Africa was still connected to Asia at this time by landbridges across Arabia and into Iran. By around 17 Ma the Red Sea basin had rifted sufficiently to connect to the Mediterranean Sea by way of the Gulf of Suez, but a landbridge continued to exist between Ethiopia and Arabia (Coleman, 1974; Crossley et al., 1992; Whybrow, 1984). The presence of large and diverse numbers of planktonic foraminiferal fauna within Early Miocene deposits at the northern end of the Red Sea basin and in the Gulf of Suez is taken to indicate a wide connection with the Mediterranean Sea, leading to increased depth of water within the basin (Ouda et al., 2000). This connection did not always remain fully open, however, and northern Red Sea and central-southern Gulf of Suez sediments from the Middle Miocene record evaporite deposits. By the Late Miocene all connection between the Red and Mediterranean Seas had ceased with the Gulf of Suez becoming integrated into the Red Sea system, but some water is believed to still have entered the Red Sea intermittently from the Indian Ocean (Ouda et al., 2000).

The start of the Pliocene saw the initiation of tectonic movement within Arabia causing the peninsula to move anticlockwise away from Africa. During the Early Pliocene the Ethiopian-Yemen-Aden isthmus was breached allowing the Indian Ocean to mix freely with the Red Sea through the Straits of Bab el Mandeb, as evidenced by Indo-Pacific fauna. This brought to an end the longstanding land connection between Africa and Southern Arabia that had resulted in much faunal migration between the regions (Coleman, 1974; Delany, 1989; Ouda et al., 2000; Whybrow, 1984). Currently the Red Sea coasts are 175 km apart at the top of the basin with the widest point near Jizan (16° 55' N, 42° 35' E) where the distance reaches 350km, however, the Straits of Bab el Mandeb, at the south of the basin, are only 30 – 40 km wide (Chapman, 1978a).

Lithological and microfacies changes within the Marsa Alam area of the Egyptian Red Sea Coast (26° 00' to 24° 50' N) have been examined in order to understand the depositional environment of the sediments and thus shed light on climatic and sea level changes within the Pliocene and Early Pleistocene (El-Asmar and Abdel-Fattah, 2000). Fossiliferous conglomerates from the Gabir Formation (Early to Middle-Late Pliocene) appear to indicate deposition under full marine conditions, although the sea-level would only have been subtidal to intertidal, with a warm-humid climate. Other facies within the Gabir Formation, including sandstones and limestones, support this conclusion, as do deposits from the Late Pliocene Shagra Formation, further confirming the opening of the Straits of Bab el Mandeb and the incursion of the Indian Ocean (El-Asmar and Abdel-Fattah, 2000). However, subsidence of the basin floor continued as represented by areas of particularly thick carbonate deposits. These deposits are often over 3 km deep, meaning that the amount of subsidence had to have been in the region of 2 km since the Late Miocene (Crossley et al., 1992; Ouda et al., 2000). The Plio-Pleistocene boundary is represented by the deposition of the Samadai Formation, which rests unconformably on the Shagra Formation. This hiatus in deposition is due to a period of uplifting in the Red Sea basin that resulted in a prominent fall in sea level. The Samadai Formation formed under extremely shallow subtidal conditions, less than 10 metres at times, with a generally dry climate, although short humid intervals are also represented (El-Asmar and Abdel-Fattah, 2000).

The above evidence for deposition of sediment under extremely low sea depths as well as periods of both subsidence and uplifting within the Red Sea Basin suggest that under certain circumstances in the Late Pliocene and Early Pleistocene the water barrier between Africa and Arabia would have been seriously compromised. Before the worldwide rise in sea levels in the Late Holocene, due to the end of the Ice Age, sea levels were at least 30 m lower than at present. There is evidence for much greater falls in depth during the Northern Hemisphere glacial maxima that occurred sometime after 2.5 Ma (Delany, 1989; Shackleton et al., 1984). Relating this to the current depths and basin composition in the Red Sea, sea levels would have to drop about 140 m in the region to the west of Al Hanish al Kabir Island, or around 174 m in the Straits of Bab el Mandeb, for dry land to connect Arabia and Africa. Much smaller drops of between 80 and 120 m are all that would be required for land with the occasional water channel to connect the two areas (Delany, 1989).

The PRISM2 dataset, that illustrates a timeslab of 3.29 – 2.97 Ma, used a sea level estimate of 25 m above modern-day sea levels based on the massive reduction of Greenland and Antarctic ice volume at that time (Crowley, 1996; Dowsett et al., 1999). However, by 2.37 Ma the first glacial maximum was well under way, with examination of benthonic foraminifera suggesting that the level of ice volume was similar to that achieved during glacial maxima in the Middle Pleistocene (Shackleton et al., 1984). As a result, based on sea level estimates for glacial maxima during the Later Pleistocene (Delany, 1989), sea levels within the Red Sea during the Late Pliocene could have been low enough at intermittent intervals to allow mammal migration between Africa and Arabia. Unfortunately, due to the scale used within the model simulations, it is not possible to reflect this situation accurately and so from the model diagrams it would appear that the land connection between these two areas is more established than is likely to be the case.

7.6 Zagros and Caucasus Mountains

As was indicated by the distribution of sites across Central Asia, for various reasons scant research has been conducted in this vast area. As a consequence little specific information is available regarding the geographic evolution of the region during the Late Pliocene and Pleistocene, although certain generalisations can be made.

The Zagros mountains, comprising parallel limestone ridges with deep valleys, run from Azerbaijan in the North West in a southeasterly direction along the Iranian border for 1600 km to the Iranian Gulf and the Straits of Hormuz (Smith, 1986). The highest part of the mountain chain is in the region of North West Iran, where the peak of Kūh-e Zard reaches 14,924 ft. The mountains developed as a direct result of Red Sea crustal spreading and the northward movement of Arabia into Eurasia (Shahar, 1994; Smith, 1986). The mountains form part of the Mesozoic-Tertiary orogeny that was also responsible for much of the Alpine orogeny in Europe. Whilst initial movement commenced in the Late Cretaceous, the bulk of the movement was not completed until the Mio-Pliocene with Plio-Pleistocene sediments deposited after the uplift (Alavi, 1994; Falcon, 1974; Ilhan, 1974).

As the movement of Arabia to the north got underway during the Tertiary, the resulting compression in Central Asia resulted in continental to oceanic subduction in the Zagros-Makran region. The concomitant development of the Zagros thrust belt, representing the boundary between the Arabian and Eurasian plates, led to the formation of the continental mountain belt in the same region (Bird, 1978; Ravaut et al., 1997). The Zagros belt consists of three parallel tectonic features, but the main mechanism through which the mountains were built was thrusting. The thrust systems moved rocks for tens or hundreds of kilometres in a N.E. – S.W. direction (Alavi, 1994). As the thrusting was repeated on a frequent basis, complex structures of superimposed rocks developed leading to thickening of the continental crust by something in the region of 10 – 15 km (Alavi, 1994), and vertical displacement of several thousand metres (Ilhan, 1974).

The main thrust zone has not remained stable throughout the development of the Zagros Mountains, with the intensity of deformation substantially higher in regions further to the N.E. of the Persian Gulf. This would suggest a southwesterly drift of the main thrust zone and has led to differential uplift of the mountain chain with the North-Eastern part of the range being higher than the Persian Gulf (Alavi, 1994; Falcon, 1974). This orogenic zone is still seismically active with some regional uplift occurring, probably as a result of isostasy (Falcon, 1974; Ilhan, 1974). However, it is clear that the majority of the mountain building was completed by the Middle Pliocene, if not considerably earlier, and as such the Zagros chain would have represented something of a barrier for migrating plants and animals in the Late Pliocene and Pleistocene (Tchernov, 1992).

The Plio-Pleistocene history of the Greater Caucasus region, between the Caspian and Black Seas, is poorly known due to the Communist party politics of the former USSR and recent regional conflicts and political instability. As a result, the timing and rate of uplift of the Caucasus Mountains is ill-defined, with various studies relying on relative dating and correlation mechanisms involving volcanism, glaciations and regional stratigraphy (Mitchell and Westaway, 1999). Recent K-Ar studies of volcanic deposits within Armenia have led to a reinterpretation of the local chronology in the area, with many deposits being assigned a younger age than previously recognised. A consequence of this is that the presumed rate of uplift of the Caucasus Mountains has also altered, with the Eastern flanks of the Greater Caucasus now believed to have been uplifted at an average rate of 0.6 mm $a^{(-1)}$, whilst the Lesser Caucasus have been uplifted at a slower average rate of 0.3 mm $a^{(-1)}$ for the past million years (Mitchell and Westaway, 1999). This recent, rapid rate of uplift is too great to be explained purely by plate collision, and so crustal thickening and magmatic activity, that commenced about 1.5 Ma, are also thought to play a large part (Gazis et al., 1995; Mitchell and Westaway, 1999).

Despite the rapid rate of uplift of the Caucasus in the Pleistocene, it is clear from palynological analysis within Western Georgia that the relief was already marked by the Middle Miocene (Stuchlik and Shatilova, 1987). Showing a typical altitudinal zonation, the lower and middle mountain belts comprised subtropical trees and shrubs, including *Moraceae*, *Symplocos*, *Mastixia* etc. Plants preferring a warm-moderate climate such as *Platanus*, *Juglans*, *Pterocarya* and *Platycarya* constituted the next belt, whilst the vegetation at the top of the mountains was characterised by cold resistant plants (*Betula*, *Carpinus*, *Fagus*) and conifers (Stuchlik and Shatilova, 1987).

Traditionally, the Late Miocene/Early Pliocene was seen as the time of major uplift within the Caucasus with an attendant change in vegetation with coniferous plants playing an increasingly important role (Stuchlik and Shatilova, 1987). However, as stated previously, the age of many of the deposits was based on regional correlation and recent K-Ar dating would suggest that many of the dates should be revised upwards with the main period of uplift and associated faunal changes occurring in the Late Pliocene and Pleistocene (Mitchell and Westaway, 1999). Nevertheless, the palynological analysis does suggest that Western Georgia acted as a refugium for thermomoderate plants, with the mountainous relief of the Caucasus acting as a barrier preventing the colder, continental climate from flooding into the basin, at least until the climatic conditions markedly worsened in the Later Pleistocene (Gabunia et al., 2000a; Stuchlik and Shatilova, 1987). However, as was indicated when determining the height of the Tibetan Plateau, using flora to deduce altitude is fraught with problems. If the recent cooling of the earth were not taken into account, the altitude indicated by the plants would be inaccurate (Molnar and England, 1990). This would not appear to be of consequence here, though, as the exact altitude is not being determined, only the fact that some zonation of vegetation reflecting relief is apparent from the Middle Miocene onwards (Stuchlik and Shatilova, 1987).

7.7 Conclusions

The foregoing discussion has shown that in many respects the Late Pliocene world was very different from the one that we see around us today and as such simplistic pictures of migration patterns are insufficient. Although the major mountain chains of Eurasia were in evidence by the Late Pliocene, they had not reached their current altitude and as a result their impact on the geography of the continent in terms of rainshadow effects or impassable barriers was reduced. The only major feature to have influenced the Plio-Pleistocene geography as much as, if not more than, today was the Tibetan Plateau. There would appear to be little doubt that the plateau had reached its maximum elevation sometime before the timescale of this research, and its impact on Eastern Asia would have been extensive.

The changing waterways and coastlines of North-Eastern Africa and Western Arabia would have been an important factor in regard to faunal and floral movements in the area. The River Nile should not be seen as a 'passageway' out of Africa during this time period, since its current route did not develop until much more recently. This may mean that the Levantine area played a less significant role for migrants than is currently believed. A more interesting point is the development of the Red Sea during this interval. The closeness of Ethiopia and the Arabian Peninsula has always created interest in terms of a route into Eurasia, but the rifting early in the Pliocene was thought to have closed this door for later emigrants. The discovery of uplifting as well as subsidence within the basin and the reduction in sea level as a result of the onset of the Northern Hemisphere glaciation have again made this direction attractive, especially with the likely starting point of the hominid émigrés being in Eastern Africa.

Another meaningful point that research into the Plio-Pleistocene climate has shown is that the monsoon system in Asia was well developed before any hominid expansion out of Africa, and would have affected any biological inhabitants of the area from 8 Ma and especially from 2.5 Ma onwards when the system strengthened markedly (Jule Xiao and Zhisheng An, 1999; Murakami, 1987). This would have meant that the local plants and animals had to adapt to a highly seasonal climate with cool/cold and arid winters and warm, humid summers. A side-effect of the monsoon, that is of huge importance to modern-day researchers, is that the vast quantities of rain in the area resulted in the devastation of sediment accumulations and the destruction of much important material that could assist in interpreting the fauna and associated flora of the region.

This description of Plio-Pleistocene geography has been of importance for illustrating the options that were available in moving out of Africa, and onwards across Eurasia, during the Plio-Pleistocene period. Whilst barriers such as the Himalayas and Zagros-Caucasus mountain chains were present, they were unlikely to have been impenetrable and so a greater amount of freedom of movement is envisaged. Similarly, the occasional relaxation of the water barrier between Ethiopia and Arabia would have increased the alternatives available to migrants. Together, these illustrate the point that there was no major geographical barrier to prevent free movement into and out of the African continent on the Plio-Pleistocene boundary and before. In GIS terminology, therefore, the palaeogeography of the research area represented an almost 'frictionless' surface and as such, in the generation of the GIS predictive model, the land can be treated as an isotropic surface.

Chapter 8

Late Pliocene and Early Pleistocene Climate and associated impact on Mammal Migration

As well as geographical changes, the research area also experienced dramatic climatic change in the Late Pliocene and Early Pleistocene that resulted in the vegetational pattern that is familiar today. It is crucial to the understanding of possible migration routes to realise how different the climate and environment inhabited by Plio-Pleistocene plants and animals were, since any theory based on modern vegetation distribution would be erroneous. (See Illustration 5 for the location of sites mentioned within the text).

8.1 The Sahara Desert

Currently, the Sahara Desert covers an area of over 8 million km^2 (approaching the size of Brazil), between 16° and 32° N, with rainfall not exceeding 100 ± 50 mm per year (Le Houérou, 1997). Geological evidence would suggest that this represents the largest extent of desert within North Africa since the Upper Triassic. Whilst an extremely arid centre appears to have been present since the beginning of the Quaternary, (1.7 Ma), or even as early as the Late Pliocene, the limits of the desert have varied quite extensively (Le Houérou, 1997; Street and Gasse, 1981). It is this variation in the extent of the margins, especially towards the east, that is critically important to this research. Although the Sahara and adjoining deserts of the Sinai and Saudi Arabia currently isolate the Asian and sub-Saharan grasslands, there may have been corridors between the two within the time-span of human evolution. The duration and extent of such passageways are of obvious relevance to questions of when – and if – Pliocene or Early Pleistocene hominids migrated out of Africa. Whilst these channels are not yet documented in detail, information from elsewhere around the margins of the Sahara, as well as in North and North West Africa, do help indicate the desert's likely extent.

The African continent started its inexorable movement northwards from the South Pole some 400 Ma ago. For considerable lengths of time the area of the modern-day Sahara was covered by tropical sea, with the sediments suggesting a tropical, subhumid environment. The current geographical location was not attained until the Neogene, *c.* 10 – 20 Ma, and even now some northward movement is still occurring (Le Houérou, 1997).

Uplift of the Tibetan Plateau and surrounding Himalayas has had a drastic effect on the climate in both the Arabian Peninsula and North East Africa from the Late Pliocene onwards. The pattern of wind circulation altered as a result of the establishment of the high altitude barrier and the easterly jet stream developed flowing from Tibet into Africa. This resulted in excessive aridity in a widespread area, covering Pakistan, Arabia, Somalia, Ethiopia and the Sahara, due to the dry, subsiding air (Dupont and Leroy, 1995; Rognon and Williams, 1977 cited in Williams et al., 1998; Singh, 1988; Williams et al., 1998). Therefore, in order to understand fully the development of aridity and desertification within the Sahara, a working knowledge of the uplift of the Himalayan-Karakorum mountains and Tibetan plateau, 2 – 3000 miles to the east, must be obtained (see Section 7).

As a consequence of the present-day climate, little fieldwork has been undertaken within the arid centre, but research on the fringes has made interpretation of the palaeoclimate of the region more accurate. Some of the problems of reconstructing palaeoenvironments within the desert include a generally low level of pollen production that is often exacerbated by poor preservation, contamination of samples by pollen from both the Mediterranean and Sahel regions and reworking of older deposits (Beucher, 1975 cited in Street and Gasse, 1981; Cour and Duzer, 1976 cited in Street and Gasse, 1981). Also, periods of higher rainfall are better documented than more arid intervals due to the preservation of more flora, fauna and sediments (Le Houérou, 1997). Furthermore, the lack of adequately dated sediments means that only a very general pattern of climate change is apparent (Street and Gasse, 1981). Nevertheless, this has not stopped researchers working in regions as widespread as the Mediterranean and Red Sea coasts, the Atlas piedmont, Chad – Tibesti, the Hoggar and the Kurkur Oasis. The picture illustrated is one of a moister and more vegetated central Saharan region during the Plio-Pleistocene, although evidence for desert/steppe is never completely absent (Busche, 2001; Street and Gasse, 1981 and references therein), however, the frequency and duration of these moister episodes still needs to be accurately dated.

Early to Middle Pleistocene deposits along the proto-Nile and its tributaries, found on the Kom Ombo Plain, preserve evidence of several pluvial periods that allowed the formation of wadi terraces. This suggests a high and occasionally torrential river discharge, if only seasonal in some cases, as a result of rains occurring with greater frequency and intensity than in modern times. The amount of vegetation was limited, however, with erosion occurring during the prolonged arid periods from which the pluvials only offered brief respites (Butzer et al., 1968).

Late Miocene to Early Pleistocene continental Quaternary deposits have been found on the western coast of the Red Sea, south of Quseir (26° to 24° 30' N) and sedimentological and magnetostratigraphic study of the deposits has been undertaken. Several different stratigraphic formations and members were identified, but only the Shagara Formation falls within the relevant timescale, dating from 2.5 – 0.6 Ma (Freytet et al., 1994; Lean et al., 1998). A reconstruction of the palaeogeography and palaeoclimatology of the coast was undertaken on the basis of several sections across fluviatile formations and alluvial cones. Several continental carbonate facies were recorded including limestone-cemented sandstones, micritic, travertinous limestones and vertical, calcareous root coatings as well as fringing reefs (Freytet et al., 1994). The commencement of Northern Hemisphere glaciations at ≈ 2.4 Ma is likely to have caused the rapid changes in deposition seen within the Shagara Formation from reef to alluvial fan, with the alluvium prograding during the wettest periods (Lean et al., 1998; Shackleton et al., 1984). Under the present hyperarid conditions the formation of carbonate is extremely limited, since for carbonates to form the climate needs to be relatively humid. Therefore, the deposition of the fluviatile formations and alluvial cones must have occurred under a climate that was at least occasionally humid with a certain amount of free standing water (Freytet et al., 1994).

Evidence for high levels of precipitation during the Early Quaternary can also be obtained from analysis of fossil landslides within the central arid core of the Sahara Desert (Busche, 2001). The landslides occurred in locations where sandstone plateaux overlay clay- and siltstone deposits. Under wetter conditions than the current arid climate, the sandstones were able to absorb sufficient water to thoroughly wet the clay- and siltstone deposits making them plastic and unstable, resulting in landslides up to 3 km wide. These landslides affect a massive area from the Atlantic Ocean in the west to the Nile River in the east and from the Mediterranean in the north to the Sahel in the south (Busche, 2001). Unfortunately, it is not possible to date accurately the formation of these deposits due to a lack of volcanic rocks in the vicinity of the landslides and the limitations of the dating range of thermoluminescence dating. However, morphological and circumstantial evidence indicate that the Sahara experienced its last particularly humid interval during the Early Quaternary, inferring that the landslides also date to this period (Busche, 2001).

Due to the severe erosion that occurred during the dry periods, resulting in the loss of sediment that had been deposited during the more humid, wet intervals, sedimentary deposits earlier than the Late Acheulian are scarce within the Saharan region (Le Houérou, 1997; Wendorf and Schild, 1980). Some Plateau and Wadi tufas dating to the Plio-Pleistocene have been discovered in the Kurkur Oasis within the Libyan Desert, Southern Egypt (23° 54' N, 32° 19' E) and analysis has shed light on the vegetation within the region (Van Campo et al., 1968; Wendorf and Schild, 1980).

The Plateau tufa can be found outcropping extensively around the Kurkur Oasis and is thought to have resulted from a long period of stream and spring deposition. The edge of the deposit is crisply defined along what is possibly an ancient drainage system (Butzer et al., 1968). The Wadi tufa is believed to have formed as a consequence of extensive erosion of the uplands as a result of water springs, implicitly indicating heavy rain, dense vegetation and a high water table (Butzer et al., 1968; Wendorf and Schild, 1980). The Pleistocene gravels found in the Kurkur area are poorly sorted and lack much stratigraphy, further reinforcing the conviction that the area was subjected to periodic torrential rain, although there is also the suggestion that there was little vegetation to halt sediment movement (Butzer et al., 1968). Whilst it is not possible to ascertain whether the Plateau tufa dates to the final Pliocene or early Pleistocene, it does represent the oldest Plio-Pleistocene sediment within the Western Desert (Wendorf and Schild, 1980). The Wadi tufa is somewhat later, being of a Middle Pleistocene age (Van Campo et al., 1968).

Several samples were taken from the Plio-Pleistocene and Pleistocene sediments at the Kurkur Oasis for pollen analysis; regrettably, none of the Plateau tufas had yielded macrofossils or stone tools (Wendorf and Schild, 1980). Unfortunately, only three of the samples were studied due to the paucity of pollen within the deposit. These samples were U-8, extracted from the Upper Member of the Plateau Tufa within the central wadi, U-4, again from the Upper Member of the Plateau Tufa within the false wadi, and U-16 extracted from Wadi Tufa I within the false wadi (Van Campo et al., 1968).

The sediments and pollen had been deposited from a local source area, seen in both the lack of equatorial fern spores and geomorphological analysis that showed the deposit originating from a watershed near to the modern oasis (Van Campo et al., 1968). Long-distance transport of pollen grains and spores by wind was an influence that could be discounted for these samples since the closest alternative sources of 'Mediterranean montane' and 'Tertiary boreal' (wooded zone consisting primarily of conifers (Luck, 1999)) pollen lay over 850 km and 1400 km away respectively. The prevailing wind was more likely to have carried xerophytic pollen from the Libyan Desert than pollen from such distances (Van Campo et al., 1968). The pollen does not appear to have been contaminated by underlying deposits since there was a hiatus in deposition between the Eocene, when the area was a marine environment, and the Plio-Pleistocene, making any transgressing pollen grains immediately apparent. Whilst there was no such obvious difference between the Plateau and Wadi tufas, the derived pollen spectra were distinct, meaning that contamination was unlikely. Therefore, it seems probable that the pollen spectra are representative of the local pollen at Kurkur during the Plio-Pleistocene (Van Campo et al., 1968).

As is fairly typical in palynology, it was not possible to classify much of the pollen to species and occasionally even generic assignments were unattainable. Due to the large amounts of pollen preserved within the samples, however, a general understanding of the local vegetational characteristics was achieved (Van Campo et al., 1968). A comparison of spectra U-8 and U-4 showed that during the deposition of the Upper Plateau Tufa the vegetation experienced substantial variation, which is likely to be a reflection of a changing climate. Within sample U-8, 16% of the pollen was attributable to relict Tertiary flora with boreal characteristics whilst 22.5% represented lowland Mediterranean elements, indicating as a minimum some local Tertiary forest with substantial precipitation (Van Campo et al., 1968; Wendorf and Schild, 1980). By the time of sample U-4, however, an episode of aridity had taken hold with a massive increase in xerophytic flora from 18% to 42.5% and a corresponding reduction in both boreal and Mediterranean arboreal elements showing that the vegetation was undergoing change. The mid Pleistocene sample, U-16, showed a further increase of xerophytic plants to 53%, although a small proportion of boreal or montane and lowland Mediterranean elements remained, indicating that the pluvial periods, relatively common in the late Tertiary, had not completely ceased (Van Campo et al., 1968).

The pollen and geological evidence from the north eastern area of the Sahara would appear to indicate considerable variability in the climate from the Late Pliocene period until the Middle Pleistocene. The Plio-Pleistocene boundary was characterised by a relatively moist climate supporting the growth of boreal forest within Egypt and the formation of carbonate sediments. On the other hand, the Middle Pleistocene was generally represented by arid-loving flora, but the formation of the Wadi tufa sediment denotes the presence of water, occasionally in large quantities, suggesting seasonal discharge (Butzer et al., 1968; Wendorf and Schild, 1980). Without doubt, there is no suggestion that the area was subjected to the continuous hyperaridity prevalent in modern times.

The Tunisian area of North Africa underwent an era of aridification at the beginning of the Villafranchian, on the Plio-Pleistocene boundary. The deteriorating tropical conditions led to the disappearance of some archaic Neogene fauna such as *Elephas africanavus*, *Stylohipparion*, *Libytherium* and *Machairodus* (Le Houérou, 1997). Whilst much of the rest of the fauna could still have been found in the area prior to human interference and destruction in the last century, the presence of *Ursus arctos*, *Rhinoceros mercki* and *Bos primigenius* indicated a cooler climate than would have been expected. In general, a shift from tropical Pliocene (21%) to Mediterranean (52%) and boreal (21%) floral species is represented in the Villafranchian in this area (Le Houérou, 1997). The flora therefore confirms the pattern shown by the fauna, that of a vegetation similar to the humid Mediterranean flora found in the region today, although cooler temperatures are indicated at higher altitudes on the mountains with *Alnus*, *Corylus*, *Tilia*, *Juglans*, *Betula*, *Taxus* and *Cedrus* all represented (Le Houérou, 1997).

Deep sea cores provide another mechanism through which the developing aridification of the Saharan region can be examined, particularly when used in conjunction with a land-based record. Continental sediments found in the Cape Bojador region of Northwestern Africa preserve a record of Quaternary climatic change. Whilst a detailed land record only extends back 30,000 years, some earlier climatic reconstructions can be attempted (Arthur et al., 1979). Lower Pliocene Neogene deposits in the Northwestern Sahara consist of widespread lacustrine deposits indicating a generally humid climate. Thick fresh water deposits recovered from the Atlas and Hoggar Mountains, probably dating to the Upper Pliocene, are another sign of a humid, subtropical environment. These sediments are then covered by carbonate and evaporite crusts in the early Villafranchian, showing that major aridification had commenced (Conrad, 1969 cited in Arthur et al., 1979). Some pluvial lakes still continued to occur, but these sedimentary layers tended to be interspersed with calcareous and ferrugenous crusts and other desert deposits (Conrad, 1969 cited in Arthur et al., 1979). It would therefore appear from the land record that for most of the Neogene humid and subtropical conditions dominated in the Northwestern Saharan region. It was only in the Middle and Late Quaternary that desert sands and aridity moved south to latitudes 10° to 14° N, and into Mauritania and Senegal (Michel, 1973 cited in Arthur et al., 1979).

Offshore from the Cape Bojador region a series of drilling cores (CORC 15–1, Spansah 51A–1, DSDP Sites 369, 397, 137, 138, 139 & 140) have provided information on the sedimentary history of the continental shelf and adjacent land-mass (Arthur et al., 1979). The continental climate and the amount of vegetational cover are among several factors that affect the quantity and type of sediment reaching the continental shelf, meaning that the sediment deposited directly reflects the environmental situation on land. This also implies that changes in humidity and aridity on the continent are reflected in alterations in both the fertility and terrigenous-volcanic input of the sediment reaching the ocean floor, so analysis of the sediment provides an indication of the climate on land (Arthur et al., 1979).

The core extracted from DSDP 397 was 1453m long, with the base of the sequence Early Cretaceous in age. It was not until nearly the top of the core, in lithostratigraphic Unit 1, that Early Pliocene and Pleistocene nannofossil ooze deposits were recorded (Arthur et al., 1979). The Pliocene and Quaternary sediments contained considerably more material from terrigenous sources than sediments from the proceeding periods; this alteration would appear to be related to significant climatic and sea-level changes. These Neogene deposits showed short-term climatic fluctuations reflecting humid/arid oscillations as evidenced by variations in the reworking of biogenic detritus, carbonate preservation and alternating deposition of clay minerals and stained quartz. Plant fibres became relatively abundant in the core during the Upper Villafranchian, probably reflecting an increase in humidity on the continent (Arthur et al., 1979). Therefore, as shown in the land-based record, this area of Northwestern Africa underwent major climatic fluctuations, from humid to arid and vice versa, during the Pliocene and Pleistocene epochs. By the time of the Upper Villafranchian, about 1 Ma, humid intervals were still occurring, suggesting that periods of hyperaridity were circumscribed.

Another deep sea core, ODP Site 658, was drilled 160 km west off Cape Blanc at 20° 45' N, 18° 35' W, a latitude currently representing the transition from Sahelian vegetation to desert (Dupont and Leroy, 1994; Dupont and Leroy, 1995; Leroy and Dupont, 1994). This 200m long record ranges from 3.7 – 1.7 Ma, although several coring gaps are apparent, and contains the best Upper Pliocene pollen record for Northwestern Africa, revealing changes in the climate and vegetation of the Sahara and Sahel regions (Dupont and Leroy, 1995; Leroy and Dupont, 1994). The sedimentation rate was relatively fast, allowing good resolution with a sample taken every 1000 to 5000 years. The sediment contained within the core originated from several sources such as river transport and wind transport from both the trade winds and the African Easterly Jet. River transport of sediment seemed to be most important from 3.7 – 3.1 Ma, although this form of deposition did continue until at least 2.2 Ma, with wind transport then taking over as the most significant source of sediment as a result of the decreased river capacity (Tiedemann et al., 1989 cited in Leroy and Dupont, 1994; Tiedemann, 1991 cited in Leroy and Dupont, 1994).

The first time period analysed ranged from 3.743 – 3.480 Ma, although documentation is incomplete due to several coring gaps. The amount of pollen entering the sample was generally extremely low, making analysis of the vegetation more difficult (Leroy and Dupont, 1994). There was a high percentage of tropical forest elements, particularly those from Sudanian and Guinean vegetation, suggesting tropical forest, wooded savanna and woodland on land, whilst percentages of Caryophyllaceae, Chenopodiaceae and Amaranthaceae (CCA) were low. The percentage value of *Rhizophora* (mangrove tree) was high enough to indicate the presence of mangrove swamps, probably growing close to the paleoriver, a position 5° further north than they are currently able to grow (Dupont and Leroy, 1995; Leroy and Dupont, 1994).

The first signs of aridification and strengthening of the trade winds were identified between 3.480 and 3.245 Ma, with percentage maxima of the trade wind indicators *Ephedra* and other 'Mediterranean' elements (Leroy and Dupont, 1994). High percentages of Cyperaceae indicated a continuing humid environment, interspersed with occasional drier periods. The first phase of severe aridity occurred at 3.26 Ma, with a maximum percentage of *Ephedra* of 4%. This stage seemed relatively short-lived, however, with tropical forest elements and humid conditions dominating once again by 3.245 Ma (Dupont and Leroy, 1995; Leroy and Dupont, 1994). This oscillation between humid and arid conditions became the established pattern until 2.609 Ma, with the climate becoming gradually drier, as shown by a decline in Cyperaceae. However, the presence of the tropical forest elements until 2.5 Ma meant that the forest belt was able to survive a considerable distance further north than it is able to today (Leroy and Dupont, 1994).

From 2.61 Ma until the end of the sequence at 1.71 Ma, the pollen record is dominated by CCA with high percentages also recorded for *Ephedra* and *Artemisia* (Dupont and Leroy, 1995). Short-term fluctuations are absent about 2.0 Ma and 1.85 Ma, although relatively common throughout the rest of this period (Leroy and Dupont, 1994). The increased percentages of CCA confirm increasing aridification and the increased strength of the trade winds is represented by a rise in percentage of 'Mediterranean' elements. After 2.1 Ma the level of pollen present within the sediment falls, probably reflecting reduced vegetation cover, and so reduced levels of pollen, within the Saharan area. At 1.98 Ma a double maximum of *Rhizophora* (10%) was recorded, indicating that once again the climate was suitable for mangrove swamps to survive at a relatively northern latitude (Leroy and Dupont, 1994). During the glacial periods of the Northern Hemisphere, the vegetation in Northwestern Africa was represented by dry elements such as CCA, *Artemisia* and *Ephedra* indicating a developing desert vegetation due to aridification and the presence of trade winds. In general, the vegetation from 2.61 to 1.71 Ma experienced latitudinal shifts from higher levels of Cyperaceae and Poaceae, representing more humid conditions and savanna, to high *Ephedra* values during the drier periods when the arid centre of the desert was able to expand southwards (Dupont and Leroy, 1995; Leroy and Dupont, 1994).

Pollen from higher up the ODP Site 658 pollen core, representing the Brunhes chron, indicated that the aridity experienced by the region in the Late Pliocene was nowhere near as severe as that occurring in the Late Pleistocene (Leroy and Dupont, 1994). Percentages of *Ephedra* and *Artemisia* dating to the Brunhes were five times higher than those of the Late Pliocene, and the percentage of 'Mediterranean' elements from the Brunhes was also considerably higher than found in the Late Pliocene. This has been interpreted to mean that the arid periods of the Brunhes chron were more severe and prolonged than those of the Late Pliocene and that the trade winds also increased in strength during this period (Dupont and Leroy, 1995; Leroy and Dupont, 1994).

Analysis of the ODP Site 658 pollen core shows that a permanent river existed in Northwestern Africa until 3.4 Ma bringing considerable amounts of sediment to the drilling site. After this date the amount of river transported pollen and sediment slowly died down, indicating a reduction in the size of the river until it apparently ceased to exist about 2.2 Ma (Leroy and Dupont, 1994). Trade wind vigour, estimated by the sum of pollen whose source area lies outside the northern Sahara and North Africa such as *Ephedra*, *Artemisia* and *Pinus*, was low for the majority of the period in question, although a continuing rise in strength was recorded. It was not until the Brunhes chron that it was powerful enough to carry significant levels of pollen from the Mediterranean area to the core site (Dupont and Leroy,

1995). Mangrove swamps occurred intermittently until 1.98 Ma, and tropical forest more continuously until 2.6 Ma, at 21° latitude, representing a northward extension of humid vegetation belts (Leroy and Dupont, 1994).

8.1.1 Summary

The evidence outlined above illustrates that the hyperaridity characteristic of Northern Africa is a relatively recent phenomenon. At least until the Late Pliocene the high mountain summits of the Sahara would have received higher rainfall than in modern periods. This rain would have ensured that the wadis continued to flow and supply water to an otherwise arid area, since low altitudinal areas may have received only little more rain than currently (Van Campo, 1975). This rainfall and associated humidity led to the northward extension of tropical forest and savanna vegetation belts and resulted in a compact core of desert, as is illustrated in the PRISM2 vegetation reconstruction (Illustration 3) (Dowsett et al., 1999).

Since the Late Pliocene it would appear that there have been at least eight major dry-wet successions within the Saharan region, with many smaller fluctuations also occurring. These oscillations are believed to be related to the glacial/interglacial cycles of the northern hemisphere, with arid periods occurring during times of glaciation and vice versa. The expansion of ice sheets at high latitudes, in both Northern and Southern Hemispheres, together with the increased height of the Himalayas, were major factors in increasing aridity. Trade winds and the easterly jet stream developed as a result of the increased height of the Himalayas and the Tibetan Plateau, but their strength was, to a certain extent, controlled by the size of the ice sheets. Large ice sheets in the polar regions resulted in steeper temperature and pressure gradients from the equator to the poles, leading to increased trade wind vigour and a decrease in precipitation (Le Houérou, 1997; Singh, 1988). Nevertheless, until at least the Early Pleistocene the wet phases occurred with enough regularity to ensure that humid vegetation never completely disappeared from the Saharan region. This implies that the central arid core of the desert would have continued to expand and contract and certainly would not have reached the extent it occupies today.

8.2 Levantine Region

The Levantine region includes the countries of Israel, Lebanon, Jordan and Syria and is about 1100 km long and 250 – 350 km wide. It represents a potentially highly important corridor for floral and faunal movements both into and out of Africa since it was, and is, the only permanently available landbridge (Bar-Yosef, 1994; Tchernov, 1987). Whilst geographically the region has changed little since the Late Pliocene, except for some earth movement along the rift valley, palaeontological and palynological analysis can help shed light on possible climatic changes within the area. There are, however, certain inherent difficulties since dating of deposits generally relies on relative techniques such as correlation of sediments or biostratigraphy as igneous deposits, needed for radiometric dating, are extremely rare (Bar-Yosef, 1994).

The most productive area of the Levant, in terms of palaeoclimatic information, is the Jordan-Dead Sea Rift that includes three basins, the Hula, Jordan and Dead Sea. Boreholes containing continuous continental sequences from the last 3.5 Ma have been drilled in the Hula and Dead Sea Valleys for natural gas and oil requirements and samples of the sediments from seven cores were available for analysis (Horowitz, 1987; Horowitz, 1989). Within Israel the Pliocene palynological deposits are divided into two zones, Pa and Pb, whilst the Quaternary sequence is split into ten palynozones, QI – QX, only the first two of which are preglacial in age dating from 2.63 – 1.99 Ma and 1.99 – 1.825 Ma respectively (Horowitz, 1988; Horowitz, 1989). Therefore, only three zones are strictly applicable to this research. All boreholes revealed generally similar patterns when the pollen spectra were analysed, suggesting that the environmental gradient currently in existence within Israel had not reached the dramatic extent visible today (Horowitz, 1987). Some differentiation is apparent, however, with the amount of arboreal pollen higher in boreholes from the Dead Sea area during zones Pa, Pb and the start of QI implying that the area was within the tropical domain and receiving more rain than more northern areas. By the end of Palynozone QI a reversal of this trend had occurred with greater quantities of arboreal pollen found in boreholes from the Hula basin indicating that more modern climatic conditions were starting to develop (Levin and Horowitz, 1987).

Analysis of 150 samples from the six boreholes showed that the thick Late Pliocene and Early Pleistocene deposits were derived from a fluvio-lacustrine environment considerably different from the modern regime. The great thickness of the QI deposit (920 m) recovered from the Melekh Sdom borehole meant that it was used as the type section, although sediments from all boreholes were employed in the investigations (Levin and Horowitz, 1987). Only the very uppermost deposits from Pb zone were analysed, but conclusions could still be drawn. The samples contained low levels of arboreal pollen, with *Picea orientalis* the most abundant species. Both steppe and desertic taxa were well represented with steppe pollen increasing in quantity further up the section, particularly in boreholes closer to the south of the region. The transition between Pb and QI zones was marked by quantities of both Juglandaceae and Betulaceae, which was an unusual combination for this date (Levin and Horowitz, 1987).

Palynozone QIa, the lowest part of zone QI, was distinguished by a large increase in the amount of arboreal pollen, although again *Picea orientalis* dominated. Zone QIb reflected a slight change in the sequence, since the level of arboreal pollen fell whilst open vegetation and desert taxa made a larger contribution to the floral assemblage. This phase was short-lived, however, as by palynozone QIc arboreal pollen, particularly *Picea orientalis*, dominated once more (Levin and Horowitz, 1987). In all three subdivisions of Palynozone QI the level of arboreal pollen was considerably greater than that present in the area during modern times, indicating that stands of spruce (*Picea orientalis*) were extremely widespread. This species currently thrives in cool, temperate climates, and that was the general climate indicated for this zone. The reduction in the amount of arboreal pollen in QIb is likely to represent a fall in precipitation levels, a factor that is also implied for the preceding Pb zone due to the low levels of arboreal pollen that were present (Levin and Horowitz, 1987).

Palynozone QII lies conformably on QI in all boreholes and can be identified by a reduction in arboreal pollen with *Picea orientalis* still dominating the small quantity present, although both oak and Juglandaceae can also be identified. *Picea orientalis* is gradually replaced by other conifers and Quercetalia as a typical modern-day Mediterranean vegetation starts to develop (Levin and Horowitz, 1987). A small amount of desert-like vegetation was also recovered from this zone, but generally steppe vegetation dominated the spectra. This has been identified as revealing a very dry, interpluvial climate with occasional thunderstorms (Horowitz, 1987; Horowitz, 1988; Levin and Horowitz, 1987).

Despite large amounts of work within the Levantine region, particularly Israel, Late Pliocene and Early Pleistocene faunal sites remain disappointingly rare (Tchernov, 1994; Tchernov et al., 1994). Although 'Ubeidiya is probably the most well-known of the sites discovered so far, Bethlehem (31° 40' N, 35° 15' E) represents the earliest mammalian faunal site within the southern Levant. It is thought to date to the Late Pliocene due to the Lower Villafranchian nature of the fossil assemblage (Bar-Yosef, 1994; Tchernov, 1988). The remains recovered included *Nyctereutes megamastoides*, *Homotherium*, *Sus* cf. *strozzi*, *Giraffa* cf. *camelopardalis*, *Leptobos*, *Gazellospira torticornis*, *Dicerorhinus etruscus*, *Hipparion* and *Elephas planifrons* (Bar-Yosef, 1994). These were generally open-country species, indicating an environment similar to an African lowland savanna, but the presence of rhinoceros did suggest that fresh water resources would also have to have been available nearby (Tchernov, 1988; Tchernov, 1994). This would certainly confirm the theory that the Mediterranean regression in the Early Pliocene had resulted in freshwater lakes remaining within the Jordan Rift Valley system (Tchernov, 1994). The absence of *Equus*, cervids and other palaeartic species could be a result of the Zagros-Caucasus mountain chain preventing southward movement of fauna (Tchernov, 1988; Tchernov, 1992), but other sites exhibiting the same faunal pattern need to be recovered before this hypothesis can be justified. The palaeoenvironmental reconstruction shows that the environment of the southern Levant in the Late Pliocene was apparently almost identical to that of East Africa. The inference would be that taxa would not have had to adapt or change in any way to survive outside the African continent at this time.

'Ubeidiya (32° 45' N, 35° E) is located 3 km south of the Sea of Galilee (Lake Kinneret) in the Jordan Valley within the Syro-African Rift. The site, with its association of stone artefacts and faunal remains, was first discovered in 1959, as a result of agricultural clearance, and much fieldwork has since taken place. The dating of the site is typically uncertain as no volcanic material is present within the archaeological deposits, although some palaeomagnetic and radiometric work has taken place in the general area together with biostratigraphical analysis (Tchernov, 1987). K-Ar dating of the basalt sequence underlying the 'Ubeidiya sediments has revealed an age of about 3 Ma. Furthermore, tectonic movements post-date the eruption of the basalt but were completed before the 'Ubeidiya sequence was deposited, indicating that the site was considerably younger than 3 Ma. Basalts that unconformably overlie sediments contemporaneous with the 'Ubeidiya site have been dated to c. 0.85 Ma providing an upper age limit for the site of about 1 Ma. This is further confirmed by the fact that palaeomagnetic analysis of deposits from within the 'Ubeidiya sequence have revealed negative polarity, and so must have been deposited before the Brunhes-Matuyama boundary (Tchernov, 1987).

The sedimentary archaeological deposits from 'Ubeidiya are over 150 m thick and have been formed into an anticlinal structure as a result of later earth movement. Four different members have been identified at the site, which from oldest to youngest are the Li, Fi, Lu and Fu Members, and they reflect the alluvial and deltaic history of the site and the changing shoreline of the nearby lake. The taphonomy of the remains has indicated that hominids played little part in accumulation of the assemblage, with fluvial processes being the most important amasser of the fossils (Bar-Yosef, 1994; Tchernov, 1987). Forty-five different mammalian genera have been recovered, including large numbers of palaeartic species, but very few from North Africa. East African fauna was also represented with species such as *Kolpochoerus olduvaiensis*, *Hippopotamus gorgops*, *Pelorovis oldowayensis* and *Crocuta crocuta* illustrating that the Levant was not completely separated from the rest of the world, although a certain amount of endemicism of the fauna was apparent, implying some isolation (Bar-Yosef, 1995; Tchernov, 1987). Biostratigraphic analysis of the species recovered, including age limits, provided further clarification of the age of the 'Ubeidiya formation. Most of the fauna appeared to be Lower Biharian (Upper Villafranchian) in age, although certain elements, particularly micromammals, did not conform to this analysis. Whilst the exact age of these faunal complexes is uncertain, a date of around 1.4 Ma is normally supported and would concur with the other dating evidence recovered for 'Ubeidiya (Tchernov, 1987).

The faunal remains that have been recovered from 'Ubeidiya contain several xerotropic and Saharan-Arabian taxa from Ethiopia (Tchernov, 1987) e.g. oryx (Bar-Yosef, 1995). This would suggest that by the later Early Pleistocene the Levant experienced a more open-country, semi-arid environment that encouraged the migration of

African elements. A more arid-adapted fauna is represented than that found at the earlier site of Bethlehem and is in agreement with the floral evidence that records a gradual increase in aridity from the Late Pliocene onwards. By 1.4 Ma the desertic belt across North Africa had grown in extent but did not completely prevent northward faunal movement, however, it did contribute to the growing differences between Eurasian and African faunas that was becoming evident at 'Ubeidiya (Tchernov, 1992; Tchernov, 1994).

The El-Ghab rift depression in Western Syria (35° 30' N, 36° 18' E), that forms part of the Levantine rift system, contains Plio-Pleistocene sediments of great depth. Faulting along the depression has led to an horst-and-graben-like structure with maximum vertical displacement in the region of 2000 m (Devyatkin et al., 1997). Deposits close to the town of Jableh, and representing the Early Pliocene, were examined in detail in order to study the palynology. The spectra revealed a domination by the arborescent-shrub group (40 – 70%), with herbaceous pollen counting for between 30 – 48% of the pollen and conifers, mostly represented by *Pinus*, comprising the remainder. This breakdown would indicate spruce-pine covered mountains with the foothill vegetation marked by coniferous and broad-leaved trees and a cool but humid sub-tropical climate typical of other Mediterranean sites (Devyatkin et al., 1997 and references therein).

Upper Pliocene and Lower Pleistocene marine deposits have also been recovered within the rift system, specifically from the Al-Kebir depression near Lathakia (35° 30' N, 36° E) (Devyatkin et al., 1996). Palynoflora obtained from Plio-Pleistocene alluvial deposits from the Jinndiriye section dating to the Upper Gilbert chron were studied. *Pinus*, *Picea*, *Quercus* and *Oleaceae* were the most common representatives of the dominant arboreal-shrub group, with other species making occasional appearances. The herbaceous group was represented by Chenopodiaceae, Compositae and Gramineae, with the spores constituting an insignificant contribution. This has been interpreted as representing a large coniferous forest in the more mountainous areas with the piedmonts having a mixture of coniferous and broad-leaved forest, whilst the plains were characterised by vegetation of a steppic nature. This would appear to reflect a more humid but cooler climate than that which exists at the present time in the area (Devyatkin et al., 1996).

The Mardido section, covering the period from 2.8 – 1.0 Ma was also examined palynologically. This sequence was over 86 m long and as a result provided evidence from several climatic episodes within this period. Initially the spectrum was dominated by arboreal pollen such as *Cedrus*, *Picea*, *Pinus* and *Abies* with other genera including *Tsuga*, *Betula* and *Fagus* also present. The herbaceous group was represented by Compositae, Gramineae and Rhodontheae (Devyatkin et al., 1996; Golovina et al., 1996). The proposed vegetational and climatic reconstructions do not differ greatly from those of the Jinndiriye section, with mixed coniferous and broad-leaved forests occurring on the more hilly regions and the plains characterised by evergreen sclerophyllous forests with some steppe vegetation, giving the impression of a wooded savanna landscape (Devyatkin et al., 1996; Golovina et al., 1996). This would indicate a warm climate with large amounts of seasonal precipitation (Golovina et al., 1996) needed to support the green vegetation.

Higher up the section there was a much greater variability in the pollen present with a preponderance of pines and herbs and other arboreal pollen at insignificant levels. This would seem to indicate an expansion of the evergreen sclerophyllous forests up the mountains, whilst the plains were covered by maquis (Mediterranean shrubby, evergreen) flora (Devyatkin et al., 1996). At the top of the section the amount of arboreal pollen increased again, although pines still dominated the assemblage. Lower altitudinal areas were once more covered by sclerophyllous forests and shrubs, with the primary vegetation on the highlands being coniferous and broad-leaved forest (Devyatkin et al., 1996). A wetter and cooler environment is implied (Golovina et al., 1996) in order to support the growth of the deciduous trees. The uppermost extreme of the section, which dates to the Middle Pleistocene (Devyatkin et al., 1996), showed the disappearance of the coniferous pollen with a large expanse of herbaceous and typical Mediterranean xerophytic pollen. This portrays an environment not dissimilar to that of today, where the level of aridity has a strong influence on the vegetation (Golovina et al., 1996).

8.2.1 Summary

Few sites containing Plio-Pleistocene faunal remains have been recovered from the Levantine area, which makes interpreting both the palaeoclimate and migration routes of animals problematic. Nevertheless, certain trends are visible. It is apparent that, along with Northern Africa, the Levant experienced increasing aridity during the period in question, although the most major change visible in the faunal assemblages was the origin of the taxa. Palaeartic species played an increasing role in the Levant that is probably related to the enlarging of the desert belts, particularly during glacial periods, that prevented taxa from African regions moving northwards.

The same general pattern of aridisation of the climate is repeated in the floral remains. The palynological results from close to the Syrian coast show climatic fluctuations during the Late Pliocene and Early Pleistocene. Initially, at about 3.4 Ma, the coniferous vegetation is seen at a maximum, with a cool and humid climate. A period of warming followed allowing the development of thermophilous plants including *Quercus*, *Carpinus* and Oleaceae. Further cooling at about 2.4 Ma, while the climate remained humid, resulted in the growth of mixed coniferous and broad-leaved forests incorporating *Tsuga*, *Fagus* and *Liquidambar* (Devyatkin et al., 1996). The end Pliocene is characterised by arid-humid climatic cycles with the arid periods discernible by a fall in the amount of coniferous and broad-leaved pollen and an increase in steppe vegetation (Devyatkin et al., 1996; Devyatkin et al., 1997; Golovina et al., 1996).

In the Jordan-Dead Sea Rift, Early Pleistocene samples show that the climate once again became more humid with increasing levels of arboreal pollen, particularly *Picea orientalis*. The arid-humid cycles continued to occur; however, it is not until Paynozone QII that the level of aridity becomes marked with a major increase in steppic and desertic pollen at the expense of arboreal pollen. This is a pattern that is also reflected in the, admittedly small, faunal evidence and was never reversed.

8.3 The Arabian Peninsula

Whilst the Red Sea separates Arabia from the African continent, in many respects Arabia should be seen as a continuation of the African scenario. The same factors that caused the hyperaridity characteristic of North Africa today also led to the development of desert conditions within Arabia during the Pliocene and Pleistocene (see above). Little research has been undertaken in the area, partially as a result of the environment, but also the political situation has kept many potential foreign researchers away. As a result, much of the information regarding the Red Sea basin rifting has been obtained from the Egyptian side of the coast (e.g. Coleman, 1974; El-Asmar and Abdel-Fattah, 2000; Freytet et al., 1994; Lean et al., 1998; Ouda et al., 2000), although research into the past climate of the peninsula has fared rather better (e.g. Al-Sayari and Zötl, 1978; Al-Sulaimi and Pitty, 1995; Delany, 1989; Jado and Zötl, 1984; Thomas, 1984; Thomas et al., 1998; Whybrow, 1984; Whybrow and McClure, 1980/1981).

In order to reconstruct the palaeoenvironment, most researchers have concentrated on sediments and geological formations from the southern and eastern continental coasts of Arabia, as well as some central Arabian areas, from where deposits containing fossilised floral and fauna have been recovered (Al-Sayari and Zötl, 1978; Al-Sulaimi and Pitty, 1995; Jado and Zötl, 1984; Thomas, 1984; Thomas et al., 1998; Whybrow, 1984; Whybrow et al., 1990; Whybrow and McClure, 1980/1981) (see Illustration 5 for site locations in Arabia). From these studies it would appear that whilst Miocene sediments are relatively common on the peninsula, (although not all contain fossilised remains), later deposits are rarer, and are infrequently studied resulting in gaps in knowledge from the Pliocene and Pleistocene periods, particularly from the western side of the peninsula. This is rather disappointing since Arabia is an important link between Africa, South West Asia and Europe (Whybrow et al., 1990) and a fuller understanding of this geographical area would be invaluable in trying to reconstruct animal migration patterns from the Miocene period onward. Nevertheless, palaeoenvironmental conclusions can be drawn from analysis of the deposits.

Currently there is no continuous river system in central southern Saudi Arabia (Hötzl, 1978), but the Arabian Shelf, which is found towards the Arabian Gulf and forms two thirds of the peninsula (Chapman, 1978a), has many fossilised wadi/river systems reflecting a time of humid periods with much surface runoff. Geological analysis of these systems resulted in them being dated to the Upper Tertiary and Lower Quaternary (Hötzl et al., 1978a), reflecting a wetter environment than is currently experienced in the area (Hötzl and Zötl, 1984).

Wadi Birk (23° N, 46° E) was part of a river system that originally covered the whole Arabian Shelf, although now it is divided into several independent basins (Hötzl and Maurin, 1978). Evidence of massive vertical erosion remains in the area, in the form of landslides and hanging valleys, caused as a result of substantial surface runoff. This precipitation originated from the crystalline rocks of the Shield area of the peninsula, which received more rainfall than the Tuwayq Mountains situated further to the east. In order for this vast amount of erosion to have occurred and the resulting sediments transported to the delta fan near Wadi As Sah'ba (23° N, 49° E), a more humid climate with substantial river capacity is required (Hötzl and Maurin, 1978). This huge delta fan, with a radius of more than 150 km, is comprised of coarse gravel with individual stones reaching a maximum of 15 cm in diameter (Hötzl et al., 1978c). Although no fossils have been found within the deltaic deposit – unsurprising in such coarse gravel – the fan has been dated to the Upper Pliocene and Lower Pleistocene on the basis of stratigraphy and the origin of the gravel material. Much of the material forming the delta had been weathered from basalt outcrops on the Shield area of the peninsula. Two of these eruptions represented possible source material for the delta and so samples were taken for potassium-argon dating. On closer inspection the younger basalt, dated to *c.* 1.1 Ma, did not reveal any obvious signs of weathering, whilst the older basalt, dated to ~3.5 Ma, was strongly eroded showing lateritic weathering that can only occur under tropical, humid conditions (Hötzl et al., 1978b; Hötzl et al., 1978c). Therefore, the erosion in the area of Wadi Birk and the subsequent transport of this material, as well as material from the Arabian Shield, to Wadi As Sah'ba had to have occurred between 3 and 1 Ma, when the climate was humid with high temperatures and rainfall, leading to much surface water (Hötzl et al., 1978c).

Important features of the landscape in the Gulf Region of the Arabian Shelf are formations of duricrust that can help shed light on climatic conditions within the Upper Pliocene. Duricrust developed on the calcareous beds of the Late Miocene/ Pliocene Hofuf Formation (Johnson, 1978) through recrystallisation of calcite within the limestone formation and also by replacing both detrital quartz and feldspar by calcite (Chapman, 1978c). In order for duricrust to form a semi-arid climate is required with annual rainfall within the region of 200 – 500 mm, although precipitation in excess of this amount would have been required to transport the gravel found within the Wadi As Sah'ba deltaic fan (Hötzl et al., 1978c). Therefore, it would appear that the Late Pliocene and Early Pleistocene was a time of fluctuating climate;

during times of considerable precipitation there was active wadi cutting, terrace formation and erosion, whilst during the semi-arid periods the duricrust formed (Chapman, 1978c).

Geological evidence for hyperaridity does not appear to be present on the peninsula until the Late Pleistocene. Within the Ar Rub' Al Khali desert, which extends almost the entire width of southern Saudi Arabia, red aeolian sand, which would be produced during extremely arid phases, is not found until 17,000 BP (McClure, 1978). This does not mean, however, that the strong surface runoff necessary for the development of the Wadi As Sah'ba deltaic fan continued for the whole of the Pliocene to Mid Pleistocene period, for if this were the case rejuvenation of the Pliocene wadi systems would have occurred (McClure, 1978). Further evidence for reduced rainfall in the Quaternary can be found in analysis of Later Pleistocene wadi sediments. Gravel deposits originating from the Shield area of the peninsula, such as are found in the Wadi As Sah'ba deltaic fan, are not seen in more recent wadi deposits where deposition consists solely of local material. This would indicate that the level of precipitation had fallen, so stopping the long distance surface water travel characteristic of earlier periods (Hötzl and Maurin, 1978).

Fossilised deposits of both fauna and flora also provide a means of interpreting past climatic conditions within the Arabian Peninsula. Whilst most faunal remains recovered from the peninsula date to either the Late Miocene or Holocene, three localities containing Early Pleistocene vertebrate remains have been discovered in North-West Saudi Arabia (Thomas et al., 1998). The remains have been recovered from lacustrine deposits found in interdune depressions within the sandy An Nafud desert and number several hundred specimens. The fossils comprised Pisces and Chelonii, although the greatest number were Mammalia and included: *Crocuta crocuta*, *Panthera* cf. *gombaszoegensis*, *Vulpes* cf. *vulpes*, Elephantidae gen. et sp. indet. (cf. *Elephas recki*), *Equus*, *Hexaprotodon* sp.?, Camelidae gen. et sp. indet., *Pelorovis* cf. *oldowayensis*, *Oryx*, Bovidae gen. et sp. indet and Alcelaphini gen. et sp. indet (Thomas et al., 1998).

Samples of enamel from the herbivore fossils recovered have been analysed in order to determine the isotopic composition of carbon found within them and assist with palaeoenvironmental reconstruction. All $\delta^{13}C$ values recorded clearly indicate a C_4 diet for the animals, implying that they inhabited open savanna, an environment further confirmed by the presence of open arid grassland/shrub loving fauna such as alcelaphines and *Oryx*; however, the presence of both fish, turtle and *Hexaprotodon* would also indicate an expanse of water nearby (Thomas et al., 1998). Whilst dating of the sediments containing the fossils proved impossible, the archaic nature of the fauna provided an Early Pleistocene limit on the age of the assemblage. The fossils show strong African connections, indicating that they arrived on the peninsula either across the Red Sea Basin during times of low sea level or via the Levantine Corridor (Thomas et al., 1998).

Late Miocene deposits from the coastal region of the Emirate of Abu Dhabi, U.A.E. have also been examined to provide evidence for palaeoenvironments. The sediments all formed part of the Baynunah Formation within the Rub' Al Khali basin, and whilst many possible sites were identified, only two, Shuwayhat and Jabal Dhanna (both 24° 10' N, 52° 30' E), were studied in detail (Whybrow et al., 1990). These sites have revealed age diagnostic fossils of *Nyanzachoerus* cf. *devauxi*, *Miotragocerus* sp., *Hexaprotodon* sp., deinotheres, gomphotheres and cercopithecines, that suggest a date of about 8 Ma, as well as a hipparionine (3-toed horse), which has a wider time range of 11.2 – 5.0 Ma. These fossils, together with other remains, point to a low energy fluvial environment complete with fresh water fish, molluscs, aquatic mammals, reptiles and wading birds. The sediments appear to indicate that the water flow experienced periods of seasonal higher energy leading to erosion of soil deposits. However, probably by Messinian times (5.5 Ma (Hsü et al., 1977)), an increase in aridity had occurred leading to evaporite conditions (Whybrow et al., 1990). It is unfortunate that this sequence did not extend to Pliocene times; nonetheless, the evidence presented does fit with the sedimentological remains of a similar age found elsewhere on the peninsula (see Al-Sayari and Zötl, 1978).

Early to mid Miocene localities along the eastern side of the Arabian Peninsula have revealed both faunal and floral evidence from which a palaeoenvironment can be constructed. The main sites in question are Jabal Dawmat al 'Awdah in Saudi Arabia (27° 12' N, 48° 10' E) and Jabal Barakah in U.A.E. (24° 6' N, 52° 28' E), although there is also a scattering of borehole samples (26° 12' N, 48° 39' E; 22° 42' N, 55° 14' E; 27° 38' N, 48° 11' E; 27° 35' N, 47° 0' E; 27° 32' N, 48° 46' E) (Whybrow and McClure, 1980/1981). Fossil mangrove root casts have been recovered from the main sites as well as vertebrate remains. Both a crocodilian and a bovid, cf. *Oioceros*, have been recovered from Jabal Barakah, whilst a mastodon, cf. *Platybelodon*, has been found at Jabal Dawmat al 'Awdah together with some shallow water marine invertebrates (Whybrow and McClure, 1980/1981). Also recovered from Jabal Barakah, within conglomerate and aeolian sand deposits, was a jaw bone and tooth from *Mastodon* (*Tetralophodon*) sp., believed to date from the Lower Pliocene (Glennie and Evamy, 1968).

These terrestrial fossils would suggest that during the Middle Miocene, as well as the Lower Pliocene, inland areas were warm with seasonal, possibly monsoonal rain, and were less humid than the tropical climate found on the tidal flats as signified by the mangroves (Whybrow and McClure, 1980/1981). This lack of aridity is also confirmed by palynological analysis of borehole deposits, with the pollen and spores showing a predominance of ferns, mosses and fresh water algae as well as large amounts of grass pollen. This leads to a vegetational reconstruction of savanna with patches of both fresh water marshland and lacustrine environments, typical of dry, but not markedly arid, tropical-subtropical climates (McClure, 1978; Whybrow and McClure, 1980/1981). This reconstruction differs slightly from that of Glennie and Evamy (1968) who interpreted the root casts not as mangroves, but as plant roots associated with low-lying swamp environments. They therefore felt that the deposit containing the *Mastodon* remains was indicative of a wadi within a desert environment. However, *Mastodon* were more normally associated with temperate climates (Glennie and Evamy, 1968), and the reconstructions proposed both by McClure (1978) and Whybrow and McClure (1980/1981) would thus appear to be more likely.

8.3.1 Summary

The evidence from geological formations and fossils would appear to show that the Late Pliocene and Early Pleistocene was the time of highest precipitation with massive erosion and transport of sediment from the Arabian Shelf. This humid phase is characterised by the presence of gravel deposits and associated drainage systems as well as extensive weathering on both Shield and Shelf deposits (Anton, 1984). By the Mid Pleistocene, however, the climate had became increasingly more arid, although pluvial periods were not uncommon, as demonstrated by restricted local water flow and the deposition of some gravel within wadi fills. The increasing aridity is confirmed by the fact that the erosion landforms of the Late Pliocene and Early Pleistocene are overlain by purely aeolian deposits that had to have formed under at least semi-arid conditions (Hötzl and Zötl, 1978).

Tectonic movements had considerably altered the Miocene land configuration by Plio-Pleistocene times, however, changes in sea level caused by both Northern Hemisphere glaciation and uplifting within the Red Sea Basin meant that at times there could have been a landbridge between East Africa and Arabia, across the Bab el Mandeb Straits. Certainly, during the Late Pliocene there was still some biotic exchanges occurring between East Africa and Arabia, as fauna associated with the African savanna has been recovered from Eurasia (Turner, 1999). Therefore, the dry semi-tropical/savanna Arabian Peninsula could have played a very important part in the migration of early hominids since a dry passageway across the Red Sea Basin would have allowed direct access out of Africa from the East African Rift Valley system without the need to travel north for hundreds of kilometres.

8.4 The Zagros and Caucasus Mountains and Central Asia

As a result of the age of the Zagros Mountain range it is likely that during the Plio-Pleistocene the surrounding area experienced a similar climate to that of the region today; with the western slopes of the mountains receiving more rain than the eastern slopes due to the rainshadow effect, and heavy snowfall on the peaks (Smith, 1986). A basic Iranian climate of wetter conditions in the west and north west with increasing aridity towards the east and south east has been suggested for the Pleistocene. More specific climatic information has not, as yet, been recovered and data from the Zagros Mountains dating to Quaternary glacial/interglacial times has been interpreted in several contradictory ways (Smith, 1986). There would appear to be no sign of large Pleistocene lakes within the centre of Iran, but many of the smaller lakes that would have been present have subsequently dried up or reduced in size. There is, however, evidence for a vast, shallow Late Pliocene lake at Kashafrud Basin, 40 – 100 km east of Mashhad near the Afghan border. Stone artefacts have also been recovered in the vicinity, but their age is unknown (Smith, 1986).

In the Lake Eğirdir region of the Taurus Mountains, Turkey (37° 47' N, 30° 30' E) a Quaternary colluvial apron has been investigated. Accumulation of colluvium is directly related to climate, particularly precipitation and temperature, although local geology also plays a part, thus allowing a tentative assessment of the Latest Pliocene and Early Pleistocene climate to be undertaken (Nemec and Kazanci, 1999). The climate in the region is believed to have been warm and humid with many lakes present in the landscape, although there were major phases of drier conditions. This is represented by multiple debris-flow deposits and rockfalls associated with heavy rain (Nemec and Kazanci, 1999). The formation of the *terra rossa* soils indicates warm temperate to subtropical climatic conditions with both humid and dry seasons (Duchaufour, 1977 cited in Nemec and Kazanci, 1999). As would be expected, this reconstruction is remarkably similar to those proposed for other countries in the region such as Israel, Syria and Arabia (see above) and would have supported vegetation markedly different from that present in modern times.

Flora and fauna recovered from the Georgian site of Dmanisi (44° 20' N, 41° 20' E), situated in the Caucasus Mountains, 85 km South West of Tbilisi, is of use when interpreting the palaeoclimate of this region of Eurasia (Gabunia and Vekua, 1995; Gabunia et al., 2000b). The site was originally investigated as part of palaeontological excavations (Bar-Yosef, 1994), but more recently hominid fragments of a calvarium, cranium, mandible and metatarsal as well as stone artefacts have been recovered (Gabunia and Vekua, 1995; Gabunia et al., 2000a; Gabunia et al., 2000b). The deposits containing the palaeontological and palynological remains are deemed to be of Late Pliocene and Earliest Pleistocene age on the basis of faunal remains typical of the Late Villafranchian period and the absolute dating of a basaltic outcrop, found 4 m below the alluvial bone-bearing deposits, to 1.85 ± 0.01 Ma (Bar-Yosef, 1994; Gabunia and Vekua, 1995; Gabunia et al., 2000a). There is controversy over the exact amount of time that elapsed between the volcanic episode and the deposition of the overlying alluvium. Analysis of the surface layer of basalt, however, reveals primary flow and cooling features with minimal erosion and no soil formation, suggesting that the timescale involved was small (Gabunia et al., 2000a; Gabunia et al., 2000b).

The fossiliferous deposit has been divided into layers on the basis of stratigraphy, although two different schemes have been used (Gabunia et al., 2000a; Gabunia et al., 2000b). Layers 5 and 4 (also known as Unit A2 and close to the basalt outcrop) contain the highest number of animal fossils, whilst most stone artefacts have been recovered from Layer 2 (Unit B1c) with the uppermost deposits appearing archaeologically sterile (Gabunia et al., 2000a; Gabunia et al., 2000b). The faunal species composition of Dmanisi is unusual since there is a preponderance of archaic fossils over their succeeding representatives, suggesting that the environment had allowed their survival, as was also implied by palynological remains discussed in the previous section (Stuchlik and Shatilova, 1987), although further work is required to confirm this point (Gabunia et al., 2000a). The presence of many faunal species at Dmanisi that are also found at widely spaced localities throughout the rest of Eurasia, such as *Soergelia*, *Panthera gombaszoegensis*, *Equus stenonis*, *Dama nesti*, *Ursus etruscus* and *Megantereon megantereon*, would suggest that the Transcaucasus region was important in terms of faunal migration (Gabunia et al., 2000a).

The faunal elements recovered from Layers 5 and 4 were substantially different from those occurring higher up the section indicating an impoverishment of the fauna that may have been linked to climatic changes (Gabunia et al., 2000a). Corresponding with the altitudinal variation in vegetation, the earliest faunal remains recovered indicate several different habitat types. Members of the forest-steppe biozone dominated the assemblage, with the abundance of cervids reflecting the forest conditions and the importance of the steppe environment indicated by the presence of *Struthio* and *Ochotona*. Microfaunal remains and *Archidiskodon* show that xerophytic conditions were present in the area in the Late Pliocene (Gabunia et al., 2000a). Several large carnivores, including *Megantereon megantereon*, *Homotherium crenatidens*, *Panthera gombaszoegensis* and *Pachycrocuta perrieri* mirror the habitat diversity surrounding the site. Palaeoenvironmental reconstructions of the region around Dmanisi at the end of the Pliocene indicate a semi-arid to warm climate (Mediterranean type) with many local water resources and a fairly open landscape with forests found on river banks and mountain slopes (Gabunia et al., 2000a).

At the beginning of the Pleistocene the Transcaucasian region, along with much of the rest of the Northern Hemisphere, experienced a general aridisation of the climate leading to an expanse of steppe at the expense of forest habitats (Gabunia et al., 2000a). Confirmation of this reconstruction is possible through analysis of the pollen grains and spores that have been recovered. Layer 5 is dominated by forest pollen, however, by Layer 4 the spectra is predominantly comprised of shrub and herb vegetation (Chenopodiaceae, *Artemisia* and *Ephedra*), with the level of arboreal pollen (*Betula* and *Pinus*) extremely low whilst xerophytic (dry-adapted) plants of the family Boraginaceae seem to have been fairly common. Vegetational variety and vertical zonality are represented by the presence of *Abies*, *Alnus*, *Ulmus* and *Salix* (Gabunia et al., 2000a).

The palaeontological site of Akhalkalaki, discovered in lacustrine deposits at the foot of Amiranis Mountain, also located within Southern Georgia, provides additional confirmation of the steppe/semi-steppe climate of the region during the Late Pliocene and Early Pleistocene (Vekua, 1986). The site was originally investigated in the 1960s and the subsequent summary of the results in 1986 dated the site to the Lower Pleistocene on the basis of the faunal remains and their similarity to remains from nearby sites (Vekua, 1986). It would appear likely, however, that this age estimate should be revised downwards on the basis of the radiometric dating of the Dmanisi basalt to 1.85 ± 0.01 Ma (Bar-Yosef, 1994; Gabunia and Vekua, 1995; Gabunia et al., 2000a). Over 4000 mammalian bones, predominantly from perissodactyla, were recovered from the site (Vekua, 1986). The carnivore remains, particularly *Panthera* cf. *tigris* and *Canis tengisii*, are typically representative of steppe and semi-steppe environments with a dry climate and this reconstruction is confirmed by the presence of species of *Equus*, *Dicerorhinus* as well as artiodactyls. Although hippopotamus remains were also identified, it is proposed that some species would have been suited to arid/steppe conditions, as opposed to humid climates, and so represent no conflict with the proposed environmental reconstruction of a steppe landscape with a dry, warm climate (Vekua, 1986).

Neogene deposits in Western Georgia near the villages of Pokveshi, Gogoreti, Chvarbeti, Tsichis Perdi and Arczeuli have also been subjected to palynological analysis in order to be able to interpret the palaeoenvironment (Stuchlik and Shatilova, 1987). Many of the Plio-Pleistocene pollen spores recovered reflect relict Tertiary vegetation that had disappeared from neighbouring regions, thus implying that the climate of Georgia was unusual when compared to the rest of Eurasia. This would appear to be due to the fact that the Caucasus Mountains prevented the worst of the increasingly arid Late Pliocene climate from reaching Western Georgia (Stuchlik and Shatilova, 1987). The early Late Pliocene deposits resembled Middle Pliocene deposits closely, with a reasonably rich pollen spectrum; impoverishment of the flora soon occurred, however, and *Pinus* became the main component. This would appear to have been a short-lived phenomenon, though, with many elements soon returning, but still slightly poorer than in earlier periods (Stuchlik and Shatilova, 1987). By around 2 Ma a moderate, humid climate was established in the area associated with the development of dark-coniferous forests characterised by the presence of *Picea*, *Abies* and *Tsuga*. Towards the end of the Pliocene there was a noticeable decrease in the numbers of subtropical and warm-temperate taxa, and warm-moderate coniferous species extended their range. This pattern continued into the Early Pleistocene, although there were increasing similarities with the modern vegetation of the area with the mesophilous plants dominating (Stuchlik and Shatilova, 1987).

Palaeoclimatological work relating to the Plio-Pleistocene period has also been undertaken along the Northern coast of the Black Sea, the South Central and South Eastern Russian Plain and further east in Tajikistan. Unfortunately, this research has failed to reveal any hominid presence, although faunal remains have been recovered (Borisova, 1994; Dodonov et al., 1999; Svetlitskaya, 1994). Some of the focus of the work has been in interpreting the loess-paleosol deposits that are widespread over Central Asia. However, the work has concentrated on certain key areas, such as Southern Tajikistan, where the loess-paleosol sections are believed to be most complete (Dodonov et al., 1999). Extending the study areas to cover more of Central Asia would be invaluable in determining climatic changes and would also lessen the necessity of attempting correlations over huge distances.

On the northern coast of the Black Sea, around the town of Odessa, continental and marine sediments have been revealed that encompass the entire Pliocene. Of the three stages of the Pliocene in this area, Pontian, Kimmerian and Kuajlnician, only the youngest, the Kuajlnician dating between 3.4 – 2.3 Ma, falls within the age range of this project (Svetlitskaya, 1994). During the Late Pliocene this region was closed and contained brackish water with the sea being present only in the south. Mixed conifer and broad-leaved trees, including *Acer*, *Tilia*, *Fagus* and *Quercus* as well as warm-loving (thermophilous) trees such as *Pterocarya*, *Zelkova*, *Fraxinus*, *Morus* and *Juglans* grew on the coast, whilst further to the east moist steppe vegetation is represented. This reflects a warm, humid climate with annual temperatures ranging from +4° C to +22° C (Borisova, 1994; Svetlitskaya, 1994).

By the end of this period the forest density had reduced considerably with pines and dark conifers dominating and the quantity of thermophilous flora had also fallen (Svetlitskaya, 1994). The most significant difference between this floral assemblage and those of the Earliest Pleistocene is the occasional presence of Myricaceae, Moraceae, Rutaceae and *Taxodium*. On the whole the assemblage indicates increasing aridity and an associated fall in temperature resulting in a more open environment with steppe vegetation (Borisova, 1994; Svetlitskaya, 1994). This is further confirmed by the absence of giraffes, *Tapirus* and *Deinotherium* that are characteristic of warm habitats and increasing numbers of more temperate faunal elements (Korotkevich, 1988 cited in Svetlitskaya, 1994).

The same picture is repeated in other parts of the Southern Russian Plain during the Late Pliocene. In the River Don basin (47° 30' N, 43° E) steppe vegetation with riparian forest was widespread at about 3.5 Ma (Borisova, 1994). A small cooling trend, reflected in a marked increase in *Betula*, occurred around 3.0 Ma, but this was short-lived with thermophilous vegetation returning. It was not until close to 2.0 Ma that the cooling led to the local extinction of exotic taxa and the expansion of steppe flora and fauna including *Equus stenonis*, *Paracamelus* and *Struthio* (Borisova, 1994). Further to the south east the vegetation was dominated by steppe elements such as *Artemisia* and *Chenopodiaceae* and coniferous forest as the area suffered from greater continentality and aridity than more western regions (Borisova, 1994).

8.4.1. Summary

The vast geographic region surrounded by the mountains of the Zagros, Pontian, Elburz, Kopet Dagh, Khurasan, Baluchistan, Taurus and Caucasus chains (Bar-Yosef, 1994) and encompassing Eastern Turkey, Iran, Iraq, Georgia and South Western areas of the Russian Federation is extremely poorly known considering its surface extent of well over 2 million km^2 (larger than the combined area of France, Germany, Italy, Spain and the U.K.) (Luck, 1999). Due to the small scale of research that has been undertaken within the Zagros and Central Asian region, mostly as a result of political difficulties, accurately assessing the part the area may have played in mammal migrations during the Plio-Pleistocene is problematic. Part of the drawback is that palaeoclimatic interpretations have to be correlated over vast distances and therefore many local differences will be overlooked as any reconstruction is of necessity very generalised. Much of the area, particularly in Iran, is uncharted in terms of Palaeolithic material, but surface collections have indicated that bifacial tools can be found anywhere. However, little else is known about the sites with dating restricted to typological assessments as the Plio-Pleistocene palaeogeography of the area is poorly understood (Bar-Yosef, 1994; Smith, 1986). Part of this paucity within Iran is related to the lack of fieldwork that has been undertaken, particularly in regions outside the Zagros area, as well as the poor quality of excavation and post-excavation work that has taken place (Smith, 1986).

Since the Zagros Mountain range had reached its modern height prior to the Plio-Pleistocene, it is likely that the climate of the surrounding area was similar to that of the region today. The main change that occurred in the palaeoclimate of Southern Central Asia during the Late Pliocene and Early Pleistocene reflects the same trend seen in both Northern Africa and Arabia at the equivalent time, that of increasing aridity (Dodonov et al., 1999). As can be observed from the loess formations of Central Asia, as well as the faunal and floral remains, the Late Pliocene climate was relatively wet and warm. The fauna was dominated by open landscape types, including savanna animals, although some taxa from forest biozones have also been recovered (Vangengeim et al., 1988 cited in Dodonov et al., 1999). Precipitation was high enough in the Pliocene and Early Pleistocene for a widespread forest zone to develop in the piedmont area of the Pamirs and southern Tien Shan (Dodonov et al., 1999) and the extensive desert areas visible today had apparently not yet developed.

Further north in Central Asia the cooling trend would appear to have commenced slightly earlier with steppe/forest vegetation characteristic of the South Eastern Russian Plain by the Late Pliocene. More north-westerly districts, around the Black Sea Coast, were still open to thermophilous flora and fauna until the very latest Pliocene, by which time the steppe vegetation had spread to cover most of the Central Asia region. This is particularly true of the Transcaucasian region of Western Georgia where the moderate height of the mountains had allowed a typical early Late Pliocene environment to continue into the Pleistocene. The presence of archaic fauna and flora, as well as more modern representatives, indicates a refugium environment from which the ecologically tolerant fauna could disperse as and when the climate allowed. This area of Eurasia would thus appear to represent a significant expanse in terms of plant and animal survival when formerly inhabited regions became unsuitable, and a place from which more adaptive taxa could migrate.

8.5 China

Interpreting Chinese palaeoclimate during the Late Pliocene and Early Pleistocene presents many problems since the various different Neogene deposits are not universally recognised and frequently their ages are unknown (Zhanxiang Qiu and Zhuding Qiu, 1995). This is due to the fact that most of the sediments are continental in origin and so often cannot be correlated with the well-known marine sequences, and volcanic deposits needed for radiometric dating are scarce. However, advances have been made in the dating of the loess-paleosol sequences on the Chinese Loess Plateau through the use of magnetic susceptibility (Huaya Lu et al., 1999; Kukla and Cílek, 1996). Nevertheless, identification and chronology is generally established through the use of fossil mammals. Comprehensive Neogene research was not initiated until the 1970's in China and still lags behind the level of knowledge achieved in Europe, although attempts have been made to collate data to enable chronologic subdivisions and correlations to be made (Zhanxiang Qiu and Zhuding Qiu, 1995).

The use of continental stratigraphy has meant that the Chinese interpretation of the Plio-Pleistocene boundary is considerably earlier than that recognised elsewhere in the world, where chronology is generally based on marine sediments as a result of their greater resolution (Zhanxiang Qiu and Zhuding Qiu, 1995). This has the potential to create considerable problems when attempting cross-continental correlations. This is especially true as some scholars working within China, appreciating the problems of having two contradictory timescales, employ the marine stratigraphic subdivisions to interpret continental facies whilst others utilise a stratigraphy based on continental deposits (e.g. Tong Guobang and Shao Shixiong, 1991 and references therein; Zeng Shaohua and Han Defen, 1991; Zhanxiang Qiu and Zhuding Qiu, 1995).

Due to the large area of China, it has long been understood that different regions of the country would have their own diagnostic fossil remains, and as a result the country has been divided into three provinces on the basis of the mammalian fauna (Xue Xiangxi and Zhang Yunxiang, 1991). The Northern China region includes the majority of Northern China from Shandong Province to Xinjiang Province and from Inner Mongolia to the first slopes of the Qinling Range (Himalayas), with most of the Quaternary fossils corresponding to the Palaeartic Realm. The Southern China region extends from the southerly aspect of the Qinling Range south to the Chinese border and from Taiwan to Sichuan and most Quaternary fossils belong to the Oriental Realm. Between these two zones, in the east of the country, is the Transitional Region (Xue Xiangxi and Zhang Yunxiang, 1991). (See Illustration 6 for the locations of the Chinese Provinces).

The various faunal units within China have been identified on the basis of mammalian classification and have generally been assigned to the Early, Middle or Late Pleistocene (Q_1, Q_2 and Q_3 respectively). In the absence of further information, such as isotopic dating or palaeomagnetism, the age of the fauna is assessed based on the percentage of surviving Neogene elements, presence or absence of typical representatives for a geological period and the proportions of extinct genera to modern (Xue Xiangxi and Zhang Yunxiang, 1991). Once this information has been obtained it would then be possible to assign the assemblage to one of several faunal stages. The material representing each faunal stage has normally been collected at a type locality, after which the stage has been named, with supplementary material obtained from locations of a similar age. Each identified stage is therefore an agglomeration of faunal remains based on mammal characteristics (Zhanxiang Qiu and Zhuding Qiu, 1995) rather than a period prescribed by chronology, and indeed the true age of each stage is open to considerable doubt.

Another problem that arises when attempting to interpret the Plio-Pleistocene climate of China is the varying interpretations regarding the rise of the Tibetan Plateau and Himalayas, as discussed in Section 7.1. Most Western scholars disagree with the traditional Chinese interpretation of recent uplift in the plateau area, but have so far failed to incorporate the revised timetable into syntheses of climate reconstructions. Chinese researchers, however, tend to intimately link the Pleistocene uplift of the Tibetan Plateau with climatic events in North China. Since the timescale to which the Chinese were working is thought to be incorrect, it is unclear how much trust should be placed in the reconstructions since the amount of floral and faunal exchange that occurred over the Himalayas is unknown. (See articles within Liu Tungsheng, 1991; Whyte, 1984; Zhang Zonghu, 1991; also An Zhisheng et al., 2001; Huaya Lu et al., 1999; Kukla and Cílek, 1996).

As a result of the obstacles inherent in the methods employed in China to construct chronological frameworks, most published palaeoclimatic interpretations, based on both flora and fauna, are of necessity very generalised. During the Pliocene, North East China was covered by subtropical forest-steppe, with pollen from *Artemisia*, Chenopodiaceae, Cupressaceae, *Populus*, *Quercus* and *Ulmus* recovered. Open country

Illustration 6. Provinces Within China.
Adapted from Li Xiwen et al., 1984: 435 & Luck, 1999: 32 - 35.

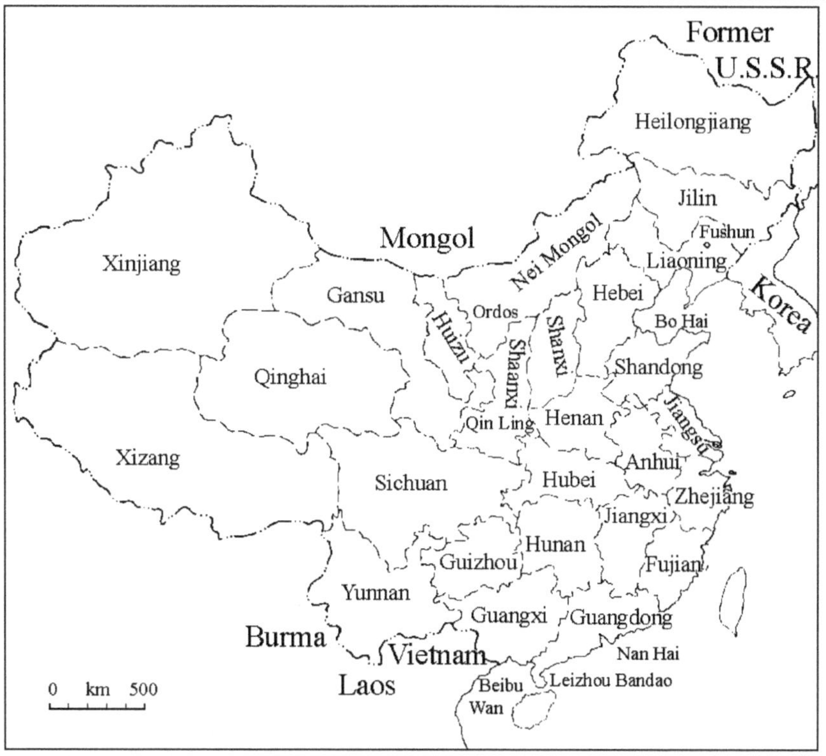

fauna including *Hipparion*, *Stegodon* and *Cervus*, although both *Stegodon* and *Cervus* were probably fairly catholic in their habitat preferences (Dennell, pers. comm. 2001), have also been found supporting this conclusion (Guo Shengqiao and Shao Shixiong, 1991; Tong Guobang and Shao Shixiong, 1991). It would seem that the more mountainous areas were covered with coniferous and broad-leafed forests whilst grasslands dominated the plains. South of the Kunlun-Qinling mountains, the climate would appear to have been warmer and more humid than further north, although the vegetation still retained subtropical grassland-forest elements with typical Oriental Realm fauna. Some cooling of the climate did occur throughout China during the Late Pliocene, a trend that continued into the Pleistocene (Guo Shengqiao and Shao Shixiong, 1991; Tong Guobang and Shao Shixiong, 1991).

On the whole, the Early Pleistocene climate in Southern China is thought to have been more stable than that to the north, with a warm, wet topical/subtropical climate continuing. The northern climate in the Early Pleistocene was cooler than that of the Pliocene, but warmer than that of the Later Pleistocene, with rain supporting numerous lakes and rivers. At the start of the period the majority of the fauna still comprised forest-living taxa, however, some grassland/steppe fauna was present in North China with *Sinomegacerus*, *Equus*, *Sivapanthera* and *Hyaena* recovered from Gonwangling, Shaanxi Province. The onset of loess deposition coincided with an increasingly cold, arid climate and the expansion of steppe flora and fauna at the expense of forest species. Subtropical taxa including *Podocarpus*, *Cedrus*, *Tsuga*, *Carya* and *Pterocarya* migrated southwards at this time (Guo Shengqiao and Shao Shixiong, 1991; Tong Guobang and Shao Shixiong, 1991; Wu Rukang and Xinzhi, 1984; Xu Ren, 1984; Xue Xiangxi and Zhang Yunxiang, 1991; Zhanxiang Qiu and Zhuding Qiu, 1995).

Some palaeoclimatic research has been conducted at individual localities, however, and this has helped provide more specific reconstructions within Central Eastern China. A 250 m thick section of loess-paleosol and fluviolacustrine sequences has been recovered from the Guanzhong Basin located in Central China, Shaanxi Province (34° 30' N, 109° 45' E). Dating by palaeomagnetism has indicated that the fluviolacustrine sediments formed between 3.08 – 1.9 Ma, and so directly relate to the Plio-Pleistocene transition (Han et al., 1997). Analysis of the grain size within the fluviolacustrine sediments revealed a strong cyclical pattern from coarse sand to silt/clay layers believed to reflect hydrological processes associated with a floodplain environment. Further environmental information was obtained from palynological analyses. Sixty samples were taken from the fluviolacustrine sequence, all of which contain pollen, although occasionally only at low concentrations. Three pollen zones were identified: 1.9 – 2.47 Ma, 2.47 – 2.7 Ma and 2.7 – 3.0 Ma; zones I – III respectively (Han et al., 1997). Stable carbon isotope analysis was also undertaken to determine the ratio between C_3 and C_4 plants and to ascertain whether the pollen data typified the palaeovegetation. A complicating factor in this analysis, discovered through modern research into stable carbon isotope values, was that values in the region of -26‰ did not necessarily indicate forest, since steppe flora is mostly composed of C_3 plants and so would return a similar value (Han, 1995 cited in Han et al., 1997).

The pollen in Zone III is predominantly non-arboreal, over 90%, with *Artemisia* forming the largest part of the assemblage. Chenopodiaceae and Gramineae are also represented with values between 10 – 30%. The few *Pinus* grains recovered were believed to have derived from the Qinling Mountains and so were not thought to be characteristic of the local flora. This reconstruction was confirmed by the $\delta^{13}C$ values of around -25‰, indicating the dominance of C_3 vegetation (particularly *Artemisia*) (Han et al., 1997). The proposed reconstruction is of a typical steppe environment with few trees and the unvarying levels of pollen attest to little environmental change. A relatively dry, stable climate was indicative of the period and the presence of elephant fossils uncovered during the fieldwork suggests warmer temperatures than are currently found in the region (Han et al., 1997).

Zone II, 2.47 – 2.7 Ma, is poorly represented in terms of pollen concentration. Once again, high levels of *Artemisia* (C_3 plants) are present, but an increase in Chenopodiaceae and/or Gramineae (C_4 plants) characterises the assemblage with few other species present, although their individual levels do fluctuate. This has been interpreted to mean that steppe flora continued to exemplify the region, but climatic variations resulted in the changes between C_3 and C_4 vegetation, as was also revealed by the $\delta^{13}C$ values (Han et al., 1997). The palynological results infer a continuation of the dry climatic regime, although the conspicuous cooling of the climate, linked to the Siberian cold high-pressure system, resulted in an impoverishment of the flora compared to the proceeding period (Han et al., 1997).

Pollen from Zone I revealed a marked change in vegetation with arboreal pollen becoming important, attaining levels of 46%. Whilst *Pinus* was the major constituent of the arboreal pollen, *Quercus*, *Juglans*, *Ulmus*, *Rhus* and *Betula* were also present. *Artemisia* was still relatively abundant, but did not reach the high levels characteristic of Zone III. The levels of arboreal pollen varied widely throughout this interval indicating major climatic alternations, with the palaeovegetation oscillating between temperate forest-grassland and arid steppe (Han et al., 1997). The sudden increase in arboreal vegetation marks the onset of wetter conditions within the region as well as an increase in temperature indicating a sub-humid climate. However, the cyclical nature of the pollen levels shows that these warm, wet periods were interspersed by cooler semi-arid/arid stages when steppe vegetation flourished (Han et al., 1997). Therefore, in this region of China, slightly North of the Himalayan (Qinling) Mountains, during the period of interest, the onset of glacial conditions would appear to have affected climatic stability, much as it did elsewhere in the research area, resulting in a climate not dissimilar from that which exists in the region today.

Not far from the previous region, in the Yushe Basin, S.E. Shanxi Province (36° N, 113° E), further research on fluviolacustrine sediments has been undertaken. The sediments studied form the Yushe Group and palynological analysis has resulted in ten pollen zones being defined. These zones encompass all four formations within the Yushe group, covering a period from 5.5 Ma – 1.8 Ma, dated through both magnetic polarity and Beryllium-10. More than 100 different types of pollen were identified within the pollen samples, although less than 5% of these formed an important part of the pollen spectrum (Shi et al., 1993).

The Early Pliocene deposits, pollen zones Y2 – Y5 (4.4 Ma – 3.0 Ma), recorded high levels of arboreal pollen, including *Pinus*, *Quercus*, *Ulmus*, *Betula* and *Alnus*, with the dominant representatives alternating between coniferous and broad-leafed species. The sub-tropical genera *Carya* and *Liquidambar* reached maximum values within this period and there were also occasional periods of herb expansion. This would indicate a warm climate with some rain (Shi et al., 1993). Pollen zones Y6 – Y8, Late Pliocene, represented a transitional period within the region. Although arboreal pollen still dominated the spectra, with *Abies* reaching its maximum level, the pollen from sub-tropical species diminished and eventually disappeared. There was a large expansion of steppe/grassland elements as well as an increase in the levels of drought-enduring *Ephedra*. The evidence would suggest that during this period the climate became cooler and drier (Shi et al., 1993). The last two pollen zones, Y9 and Y10, dated from 2.3 – 1.8 Ma, confirmed the trend towards cooler and drier conditions. The dominate taxa were herbaceous, particularly *Artemisia* and Chenopodiaceae, reflecting a marked increase in steppe vegetation indicative of a cold and dry climate (Shi et al., 1993).

Another locality within Central Eastern China where pollen has been analysed in order to reconstruct the palaeoclimate is Jing-Le Basin, Shanxi Province (38° N, 113° 30' E). Within the basin the Late Pliocene 'red clay', dated by biostratigraphy, underlies loess and overlies alluvial sand and gravels. It has been suggested that the 'red clay' represented a transition in depositional processes from dominantly alluvial to dominantly aeolian at the Plio-Pleistocene boundary (Duowen and Derbyshire, 1991).

The amount of pollen found within the 'red clay' was low, but did allow some indication of the palaeoclimate. Broad-leafed arboreal pollen was the most common constituent of the spectrum, representing 35% of all pollen grains, the most common species being *Betula*, *Quercus*, *Corylus*, *Celtis*, *Ulmus*, *Carpinus* and *Castanea*. There was also a large herbaceous element (33%) dominated by *Artemisia*, although Chenopodiaceae and Gramineae were also present. A small amount of pollen from *Pteris*, Liliaceae and Ranunculaceae was also recovered; these taxa can be found in sub-tropical habitats, but are not of themselves diagnostic of a sub-tropical climate since they can be found in other environments. The amount of pollen from both coniferous and arid-tolerant taxa was also low (Duowen and Derbyshire, 1991).

The pollen spectrum implied a warm, temperate wooded grassland habitat and was supported by the available faunal evidence of *Hipparion*, *Aceratherium*, *Stegodon*, antelope and gazelle. Further confirmation of this climate was provided through analysis of the sedimentary deposits within the basin. The 'red clay' contained numerous carbonate concretions that could only have formed under conditions that were neither extremely wet nor dry, implying that when deposited the climate was subhumid. It was thought that the climate in the area became increasingly humid resulting in the leaching of the carbonate from the 'red clay' layer soon after deposition. Drier conditions had to have been established by the Lower Pleistocene, however, enabling loess to be deposited over the 'red clay' (Duowen and Derbyshire, 1991).

Much recent research within the Plio-Pleistocene of China has focussed on the Chinese Loess Plateau, which is located to the north-east of the Tibetan Plateau. Analysis of the loess-paleosol succession, which provides a long and continuous palaeoclimatic record of the development of the East Asian monsoon, has provided new insights into the changing climate of the region and has the potential to be of more importance than palynological deposits. (See An Zhisheng et al., 2001; Ding et al., 2000; Ding et al., 1988; Ding et al., 1999; Ding and Yang, 2000; Huaya Lu et al., 1999; Junfeng Ji et al., 2001; Tungsheng Liu et al., 1999; Xiao Zhang et al., 1999). It has been estimated that loess deposition commenced on the plateau *c.* 2.4 Ma, indicating a period of increased aridity as well as an enhancement in the strength of the winter monsoon system. The paleosols developed when the summer monsoon was dominant, and are consistent with interglacial periods (Ding et al., 2000; Huaya Lu et al., 1999). These developments are thought to be linked to either an expansion in the polar ice sheets or to sudden uplift of the Tibetan Plateau area (Huaya Lu et al., 1999; Jian Wang et al., 1999; Kukla and Cílek, 1996), although as Section 7.1 discusses, an expansion of the ice sheets would appear more probable.

8.5.1 Summary

Despite the highly generalised nature of palaeoclimate reconstruction within China, the area where research has been more focussed, the provinces of Shaanxi and Shanxi, appeared to confirm the climatic trends simulated. There was a noticeable change in climate between the Pliocene and Pleistocene periods, no matter what chronological scheme was used. Temperatures fell with the onset of the glaciation in the Northern Hemisphere, a tendency that is visible elsewhere in the research area, and climatic variations occurred with more rapidity than had been seen during the Pliocene. The humid, subtropical Pliocene climate north of the Himalayas quickly became replaced with dry, arid weather with a concurrent increase in steppe vegetation during the Pleistocene. More southerly regions continued to exhibit evidence of subtropical conditions, however, providing a dramatic example of the influence of the Tibetan Plateau and associated mountain belt.

8.6 Conclusions

The evidence outlined above illustrates that the same general climatic pattern is visible over the entire research area – that is of increasing aridity, with the climate changing from one of relatively high humidity to today's generally arid climate. There is no question that the onset of glaciation, in both the Northern and Southern Hemispheres, was the driving force behind this change, but the mountain orogenies that were characteristic of the Miocene to Pleistocene periods also played a part.

The evidence obtained from continental and oceanographic pollen analysis and assessment of faunal assemblages has shown that the vegetational and climatic patterns of the Late Pliocene were not dramatically different from those proposed for the Mid Pliocene period (Barron et al., 1994; Dowsett et al., 1999). The reduced coverage of the Saharo-Arabian desert at the start of the research time period meant that the grassland vegetation belt extended from Western Africa across the entire Asian continent. This could then have acted as a 'corridor' for floral and faunal migration with genera leaving and entering Africa either through the Levant region or via the Bab el Mandeb Straits into Arabia.

The onset of the Pleistocene, however, corresponds with a marked decline in climate that becomes more apparent as the epoch progresses. This resulted in a decrease in the amount of precipitation for the vast majority of the area under consideration. As a result, the desert habitats extended their influence and it became increasingly difficult for most taxa to cross the barrier in either direction, and led to the extinction of less adaptable genera. Some areas, such as in Georgia, however, were protected from the worse extremes of the climate by local topographic conditions and in these cases archaic taxa were able to survive longer.

This analysis of Plio-Pleistocene climate has shown that a remarkably similar pattern of climate development is evident over the entire research area. At the onset of the time period under consideration there was considerably more free water about leading to an extension of more humid vegetation into areas that are currently considered desert. It would appear likely that at this time movement into and out of Africa was unrestricted by either climatic or topological barriers. Towards the end of the period, however, climate and vegetation were becoming increasingly more modern in appearance. This is likely to have reduced the ease with which floral and faunal genera were able to move around the African-Eurasian area.

Chapter 9

Indonesia and the Siwaliks of the Subcontinent

9.1 Indonesia

Whilst the majority of Indonesia remains unexplored in terms of Plio–Pleistocene deposits and associated floral and faunal remains, with no fossils recovered from Sumatra and only a few teeth found in Borneo (Braches and Shutler, 1984b), the island of Java is an exception. Interest in the Far East commenced in the 19th Century when Haeckel's 1868 *Natürliche Schöpfungsgeschichte* was published, followed by its English translation *The History of Creation* or *The Development of the Earth and its Inhabitants by the Action of Natural Causes* later in the same year (Haeckel, 1868; Lankester, 1868). Haeckel proposed a long vanished continent called Lemuria situated in the Indian Ocean between Madagascar and Indonesia. On this landmass lived anthropoid apes that migrated both east and west when they were forced to abandon the continent as it sank. Haeckel believed that the eastern migrants were hominid ancestors and named the hypothetical hominid *Pithecanthropus alalus* (Haeckel, 1868; Haeckel, 1923 cited in Dennell, 2001).

Haeckel's writings were a major influence on Dubois and were the main reason that he set off for Java in 1887 to find the 'missing link' (Bartstra, 1982; Dennell, 2001; Swisher et al., 1994). In 1890 he found a small fragment of a human mandible at Kedung Brubus, and in 1891 he made his most famous discovery of the type specimen of *Pithecanthropus erectus*, later subsumed within *Homo erectus*, at Trinil and named in honour of Haeckel (Bartstra, 1982; Dennell, 2001; Orchiston and Siesser, 1982). These fossils resulted in huge interest amongst European palaeontologists, as Dubois was able to convince them that he had found a hominid, although he did later recant this interpretation (Bartstra, 1982; Swisher et al., 1994). Further expeditions to Java were mounted in 1907 – 1908 by Madame Selenka, and in the 1930s, most notably by von Koenigswald, as well as more recently, and led to the discovery of four additional sites at the modern-day localities of Ngandong, Mojokerto, Sangiran and Sambungmacan (de Vos et al., 1982; Orchiston and Siesser, 1982). (Illustration 7 shows the locations of some of the principal sites). However, despite the history of discoveries in Java, the original provenance of much of the key fossil material and many of the intricacies of the geology remain uncertain to this day, making it difficult to date and ascertain the importance of the remains recovered.

In the 1930s Von Koenigswald provided much needed vertebrate palaeontological experience on Java, as originally Dubois had assigned the same age to all the faunal remains, making it difficult for geologists to survey the island correctly (Braches and Shutler, 1984c). Von Koenigswald was responsible for interpreting all the Pliocene and Pleistocene fauna on the island and producing a framework that, despite his reservations, was used extensively for over 50 years. He assigned the fauna into five groups that from the oldest to the youngest were: – Cijulang, Kali Glagah, Jetis, Trinil and Ngandong (Braches and Shutler, 1984c). The different faunal types were identified by the use of 'guide fossils'; a practice questioned by contemporary researchers but none-the-less supported for many years (Sondaar, 1984). Another query with Von Koenigswald's biostratigraphy was that the fauna representative of the Cijulang and Kali Glagah types consisted of only a few remains, leading to suggestions that distinguishing between the two types was inappropriate (Braches and Shutler, 1984c). Furthermore, Von Koenigswald undertook the entirety of his research from the laboratory, rarely going into the field to verify find localities. Fossils were often collected by hired workmen, packed into crates and then sent by train to the local museum where Von Koenigswald

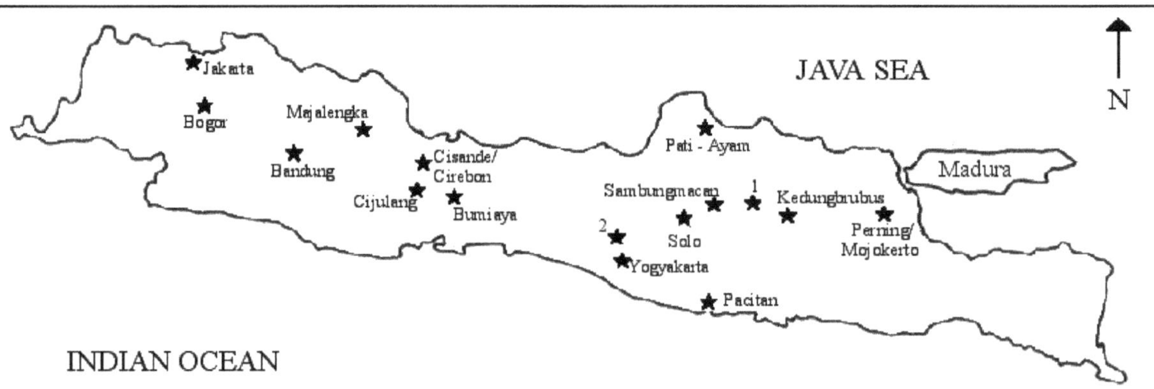

Illustration 7. Locations of Various Key Localities Within the Island of Java, Indonesia. Adapted from Sémah, 1986: 363.

unpacked them (Dennell, pers. comm. 2001). Nevertheless, it was not until the mid 1980s that reassessment of the faunal remains began (Braches and Shutler, 1984c); by which time the use of relative and absolute dating techniques had made the task easier, although not without complications.

Parts of Java became dry land in the Late Cainozoic, and there was a widespread belief that this process began in the west of the island and proceeded eastwards (Braches and Shutler, 1983). Von Koenigswald's oldest Javan fauna, represented by the Cijulang and Kali Glagah faunal types, had been recovered from close to the Cijulang River south of Cirebon and near Bumiaju respectively, localities that he identified as being in Western and Central Java. No earlier remains had been found further to the west (Braches and Shutler, 1984a). Progressively younger fauna was then to be found towards the east as a result of the later emergence of land in that area. However, more recent interpretations of the faunal remains shed doubt on Von Koenigswald's original hypothesis (Braches and Shutler, 1983). The two main problems with the theory of west to east emergence were that there was no certainty that the Cijulang fauna was actually the oldest on the island, and locality maps of Java indicate that both Cirebon and Bumiaju are centrally located (see Illustration 7). This would seem to be at conflict with Von Koenigswald's theory (Braches and Shutler, 1984a). A more likely scenario was proposed by van Bemmelen in 1949 who considered that parts of Central Java emerged as the first areas of dry land (Van Bemmelen, 1949 cited in Braches and Shutler, 1984a). This was later supported by Sondaar's 1981 paper, which suggested that certain aspects of the Kali Glagah fauna would have been undaunted by a water crossing, inferring that at one stage Java could have consisted of several small islands (Sondaar, 1981 cited in Braches and Shutler, 1984a).

Further evidence of the inaccuracy of Von Koenigswald's biostratigraphy was discovered by de Vos et al. (1982) when they interpreted the Trinil faunal type as being older than the Jetis fauna. They were forced to reassess most of the fauna classified by Von Koenigswald, since many of the faunal lists used by him to set up his biostratigraphy had come from multiple sites (Hooijer and Kurtén, 1984). Also, it was not unknown for Von Koenigswald to omit certain species from faunal lists if they did not fit with his perceived plan. They also found that they were unable to utilise the lithostratigraphy to confirm their conclusions since it had also been based on Von Koenigswald's faunal interpretations (de Vos et al., 1982). They discovered that the fauna from Trinil had many endemic species and also lacked some of the more advanced forms found at both Jetis and Kedung Brubus, thereby indicating that the Trinil fauna had to be older (de Vos et al., 1982; van den Bergh et al., 2001). These conclusions, however, were not widely accepted with both Bartstra (1983) and Hooijer and Kurtén (1984) preparing strong rebuttals.

De Vos et al. (1982) did not examine the Jetis fauna itself when undertaking comparisons with the Trinil fauna, but instead used remains collected from the Kedung Brubus area that had been regarded as an almost identical assemblage to that representing the Jetis fauna (Hooijer, 1975 cited in de Vos et al., 1982). The fauna comprising the Kedung Brubus, however, had been collected at the centre of the village of Kedung Brubus as well as up to 1 km away from this location and no differentiation between these localities had apparently been made by Dubois whilst collecting (Bartstra, 1983a; Hooijer and Kurtén, 1984). Geological studies within the area have shown that the rock beds dip to quite a substantial degree resulting in sediments of differing age being exposed at the surface. As a result, the fauna collected by Dubois and assigned to Kedung Brubus cannot be seen as a cohesive unit (Bartstra, 1983a) and so should not be used for comparison with other faunal collections.

The fact that the revised classification of de Vos et al. (1982) was based solely on biostratigraphic grounds, and ignored the changes in stratigraphy, led partly to its dismissal in certain quarters (Hooijer and Kurtén, 1984). Furthermore, it was believed that some of the arguments put forward by de Vos et al. were inaccurate, particularly as regards the influx of Asian migrants (de Vos et al., 1982; van den Bergh et al., 2001). Several of the species epitomising the Kedung Brubus fauna were actually endemics, and so did not represent an incoming faunal group, this resulted in Kedung Brubus actually having more endemic species than Trinil (Hooijer and Kurtén, 1984). However, these points did not fully address the concern of de Vos et al. (1982) that advanced forms of *Elephas*, *Rhinoceros kendengindicus*, *Epideptobos* and *Tapirus* were apparently missing from Trinil whilst present in the supposed older fauna of Kedung Brubus. It has been suggested (Hooijer and Kurtén, 1984) that these advanced forms were not absent from Trinil, but have simply been misidentified, or not identified, due to the fragmentary nature of the fossil remains. Also, the reason that they are found within the Kedung Brubus fauna was that this fauna was representative of a long time frame, so long in fact that the Trinil fauna could be encompassed within it (Bartstra, 1983a).

De Vos et al. (1982) and Sondaar (1984) were not swayed by the above criticisms and further stated that they had ensured that all faunal remains included by them within the Kedung Brubus had come from a single locality within Kedung Brubus village. There appeared to be no way out of the deadlock without resorting to dating techniques that could date the actual formations from which the remains had been derived.

Due to the complex lithostratigraphy on the island, in the 1970s and 1980s an Indonesian–Japanese CTA–41 project to define the various formations, centring on the Sangiran Dome, was initiated (Larick et al., 2000 and references therein). Five Plio–Pleistocene formations unique to the dome area were identified: Puren (formerly Kalibeng), Sangiran (Pucangan), Bapang (Kabuh) and Pohjajar (Notopuro) (Larick et al., 2000). Within the literature, however, these names tend to be used interchangeably. This confusion over Javan geology and faunal age is typical of all the deposits of a Pliocene date or younger, and is responsible for much confusion over the importance of the hominid remains.

The ages of the Pucangan and Kabuh Formations, from which the hominid remains have been recovered, as well as the Solo Terrace, have been in contention with biostratigraphical analysis only adding to the problems. Several K/Ar dates of deposits have been recorded, but only a few have been published in full and some of these have lacked standard deviations or have unacceptably high error limits (Orchiston and Siesser, 1982). Attempting to use these dates to understand the hominid remains is fraught with difficulty as many of the samples were taken from sites some distance from the hominid localities and as a result of the complex geology on the island undertaking cross correlation is problematical (Orchiston and Siesser, 1982). Added to this was the confusion over the exact locations of the hominid remains, due to them generally being excavated by local farmers and subsequently purchased by palaeontologists. Nevertheless some restrictions could be placed on the age of the different formations through the use of K/Ar analysis. The boundary between the Kalibeng and Pucangan Formations dates to between 2.5 – 3.0 Ma, the Kabuh/ Pucangan boundary to 1.2 – 1.4 Ma and the boundary between the Notopuro and Kabuh Formations to 0.69 Ma. These dates, however, appeared to be at variance with those obtained from palaeomagnetic analysis and so further research was clearly necessary (Orchiston and Siesser, 1982).

Despite the problems of accurately locating the find spots of the hominid remains, Swisher et al. (1994) undertook reinvestigations of the *H. erectus* material in order to ascertain its age. The oldest remains recovered from the Pucangan Formation were the juvenile calvarium of Mojokerto 1 and crushed facial and crania fragments from the Sangiran dome (S27 and S31) (Swisher et al., 1994). Some sediment still remained adhering to the calvarium of Mojokerto 1 but, unfortunately, not enough to perform an analysis. However, it was possible to determine that strata close to the original find location, and from which material was removed for dating, strongly resembled the pumice and volcaniclastic material in the skull. This material was then dated by means of $^{40}Ar/^{39}Ar$ analysis and an average age of 1.81 ± 0.04 Ma was obtained (Swisher et al., 1994). The closest deposit to the site of the Sangiran hominids suitable for $^{40}Ar/^{39}Ar$ dating was located in a pumice layer 2 metres above the fossil horizon and yielded an average age of 1.66 ± 0.04 Ma (Swisher et al., 1994). Further confirmation of the age of the Mojokerto 1 remains was obtained through palaeomagnetic analysis. Six samples from the site were subjected to thermal demagnetisation with a resulting normal polarity being displayed. This appeared to suggest the presence of the Olduvai subchron since this would corroborate the $^{40}Ar/^{39}Ar$ dates (Swisher et al., 1994).

These dates were not widely accepted in Europe, especially amongst the Dutch who had undertaken considerable amounts of research on the island, but met with more widespread approval in the United States. As mentioned previously, the method of excavation in Java meant that the actual find locality was passed on by word of mouth; meaning that the formation from which the remains were recovered was open to doubt and so dating the finds would always prove to be difficult (Wood and Turner, 1995). However, recent archival research in Java by Huffman (2001) has presented new evidence that suggests that the Mojokerto calvarium was recovered *in situ* from Pucangan Formation sediments.

The initial announcement of the discovery at Mojokerto was made by Von Koenigswald (Bernet Kempers, 1982 cited in Huffman, 2001), but along with other remains identified and announced by him, he was not present when the fossil remains were recovered and so was unable to describe their context (Huffman, 2001). When the first criticisms about the find were made by Dubois in 1936, Von Koenigswald was forced to quote the contextual description recording the *in situ* discovery of the fossil provided by the geologist Duyfjes who supervised the fossil extraction (Dubois, 1936 cited in Huffman, 2001). Additional corroboration of the locality was made in 1938 when visitors, including de Terra, Von Koenigswald and Teilhard de Chardin, viewed the excavation and confirmed that the stratigraphic position of the skull had been within a broad anticline. This was based on both visual inspection of the surroundings of the findspot and comparison of sediment adhering to the skull and from the site (Huffman, 2001 and references therein).

If these accounts of the locality and *in situ* discovery of the calvaria were to be believed, there still remained the task of categorically determining from what formation the fossils were recovered in order to convince the sceptics. Duyfjes was responsible for all geological mapping in the Perning/Mojokerto locality, and whilst his untimely death meant that many of his reports were unpublished, archival reports held in Bandung provided evidence for the rigorous nature of his research (Huffman, 2001). Duyfjes identified the find location as within the Pucangan Formation. He had mapped long sections of these deposits, the accuracy of which can be checked on modern air photographs and radar images, and was able to differentiate clearly between the Pucangan and overlying Kabuh Formation, a fact ignored by de Vos et al. (1982) (de Vos et al., 1982 cited in Huffman, 2001). As yet, however, the exact findspot within the Pucangan Formation has not been rediscovered and so samples obtained for dating by Swisher et al (1994) were only from approximately the same location as the fossil hominid (Huffman, 2001). Since the timescale for the deposition of the Pucangan Formation is unknown, it is not possible to say whether this represents a major problem when relating the $^{40}Ar/^{39}Ar$ dates to the fossil.

Japanese and Dutch researchers have had a long history of working on Java and previously published dates of such antiquity have been shown to be inaccurate with most researchers agreeing that, on the basis of magnetostratigraphy and fission track, the hominid remains are likely to be around 1 Ma (de Vos and Sondaar, 1994; Hyodo et al., 1993; Langbroek and Roebroeks, 2000; Matsu'ura, 1982; Pope, 1983). Although both Swisher et al. (1994) and Hyodo et al. (1993) identified a period of normal polarity at the Mojokerto site, they differed in their assignation of the section within the palaeomagnetic timescale, with Swisher et al. referring it to the Olduvai Event whilst Hyodo et al. saw it as representing the Jaramillo Event (Hyodo et al., 1993; Swisher et al., 1994). Swisher (1994) interprets this variance to be due to the fact that Hyodo et al. (1993) failed to remove the entire geomagnetic overprint as a result of only utilising alternating field demagnetisation as opposed to thermal demagnetisation in their palaeomagnetic analysis. The failure to resolve this problem and the fact that the difference represents about 1 million years is of huge magnitude when trying to interpret the migration patterns of early hominids and is part of the reason why Java remains a focus for research activity.

Further $^{40}Ar/^{39}Ar$ dates and palaeomagnetic analysis have been carried out more recently by Sémah et al. (2000). They dated the 'lower lahar', a volcanic sediment layer at the base of deposits in the Sangiran dome that marks the emergence of dry land in the area and as such circumscribes the first appearance of early hominids at the locality. The lahar from two sites, Puren and Cengklik, was analysed since the deposit varies widely in thickness and so is likely to represent longer than an instantaneous event (Sémah et al., 2000). Dates in the range of 1.66 ± 0.04 Ma and 1.77 ± 0.08 Ma were obtained for each site respectively. These dates should be used with prudence however, as the age spectrum displayed by the grains of sediment analysed appeared to suggest the presence of excess argon within the sample leading to a date that could well be slightly too old (Sémah et al., 2000). Therefore, in order to ascertain the accuracy of these dates, palaeomagnetic analysis was also undertaken on the lahar. Sediments of normal polarity were recovered from Cengklik, whilst at Puren the lahar was of negative polarity. This has been interpreted as representing the period immediately at the end of and following the Olduvai normal subchron and is consistent with the $^{40}Ar/^{39}Ar$ dates obtained (Sémah et al., 2000). A point to bear in mind, however, is that the dates obtained by Swisher et al. (1994) and Sémah et al. are almost identical, suggesting that it is possible that the American team dated the underlying lahar deposit as opposed to the Pucangan Formation (Sémah et al., 2000). Since the Pucangan Formation, from which the oldest hominid remains have been recovered, overlies the 'lower lahar', it follows that in any case the hominids cannot date to a period earlier than the Olduvai Event, *c.* 1.7 Ma. However, it is not certain how long after the lahar layer the overlying formation was deposited (Sémah et al., 2000) and so the dates obtained by Swisher et al. could still be applicable (Swisher et al., 1994) if the sample dates do relate to the hominids, but on the other hand some considerable length of time could have elapsed between the two episodes.

More compatible with the dates obtained by Sémah et al. (2000) are recent $^{40}Ar/^{39}Ar$ dates obtained by Ciochon et al. (2001) and Larick et al. (2001). They examined sediments from the South East quadrant of the Sangiran Dome where the Bapang (Kabuh) Formation has yielded almost 80 *H. erectus* specimens, over 50 of which have known findspots (Ciochon et al., 2001). The sediments at this location have a large exposure with a complete tephro–stratigraphic record and exhibit little deformation. The sequences at five different localities, containing 28 hominid fossils, were investigated with $^{40}Ar/^{39}Ar$ techniques. Dates ranged from 1.51 ± 0.08 Ma at the Bapang/Sangiran Formation boundary to 1.02 ± 0.06 Ma at the top of the hominid fossiliferous sequence (Larick et al., 2001). Three new partial hominid crania were discovered during this work and the sediments with which they were associated have been dated to *c.* 1.25 Ma (Ciochon et al., 2001). However, as the hominid occipital Brn–1996.04 was found below the Bapang/Sangiran Formation boundary, this specimen must be older than 1.51 ± 0.08 Ma (Larick et al., 2001), so early colonisation of Java cannot be ruled out.

Researchers who support a still later date for the arrival of hominids on Java are particularly interested in the tektite layer found on the island. The tektites were discovered by Von Koenigswald over 60 years ago and their presence has been well documented (Langbroek and Roebroeks, 2000). The age of the formation of these tektites, caused by the impact of a large asteroid, is well–known having been dated by K/Ar, Ar/Ar and fission–track techniques from several sites in South East Asia and Australia, and is between 700,000 – 800,000 years old (Langbroek and Roebroeks, 2000). The tektite layer is therefore to be found slightly below the Brunhes/Matuyama geomagnetic reversal (Itihara et al., 1994 cited in Langbroek and Roebroeks, 2000). There is a suggestion, however, that the tektites may have been reworked into a younger deposit, meaning that the hominid fossil layer in which they are found could be of an even more recent date (Bartstra, 1983b cited in Langbroek and Roebroeks, 2000). As a result of this data, the Sangiran hominid fossils would have to date to between 1.07 Ma and 780,000 (Langbroek and Roebroeks, 2000). The accuracy of these dates has been challenged by Swisher (1997), however, who, having carried out numerous $^{40}Ar/^{39}Ar$ dates on stratified volcanic deposits at Sangiran, suggested that all the hominid remains recovered were 1.7 – 1.0 Ma old (Swisher, 1997).

An additional factor that helps explain why the hominid remains from Java are so problematic is that there is no good evidence from the nearby Asian mainland that could indicate from where, or when, the hominids originated. It has been suggested that parts of the Sunda shelf between Malaysia, Sumatra and Borneo could have been exposed due to eustatic changes in sea level thus allowing a relatively dry passage for megafauna to reach Java, probably via the East of Sumatra (van den Bergh et al., 2001). This hypothesis has not been backed up by the available fossil record, however, signifying that further work in the area is required (Braches and Shutler, 1984b; Shutler and Braches, 1985). Indications from the mainland for early hominids is mostly lacking, with the earliest secure date for South China of *c.* 803,000 ± 3000 years obtained from the Boise Basin, in association with Acheulian–like stone tools (Hou Yamei et al., 2000). A much earlier date of *c.* 1.36 Ma has recently been proposed for sediments in association with a Palaeolithic stone assemblage in the Nihewan Basin, North China (Zhu et al., 2001), but this is still sometime after early hominids are thought to have reached Java. There is also a claim that stone flakes found in northern Thailand may date to 730,000 (Klein, 1989), but again this is well outside the proposed timescale of early colonisation. Whilst this paucity of evidence may partly be attributable to a lack of field research and the difficulties of working in densely vegetated environments, it does present an added difficulty when proposing how hominids reached Java.

9.1.1 Summary

The confusing issue of dates, with the absolute dates obtained by Swisher (1994, 1997) seemingly at odds with the rest of the age data, most notably the palaeomagnetic evidence, is the main reason why the Javan hominid fossils remain contentious when theories of hominid migration are developed. Whilst there has been no suggestion that Swisher et al. are incompetent at Ar/Ar analysis, there is much doubt over the actual deposits that they dated and until this problem is resolved there seems to be little likelihood that a date of 1.8 Ma for hominid migration to Java will be accepted. Nevertheless, this is an highly important area of research since a definitive answer could help settle many of the arguments relating to the timing of the first hominid migration.

9.2 India and Pakistan

India and Pakistan are something of a paradox, since whilst they have been intensively studied in the past, long breaks in research have also occurred leading to neglect of their important Plio-Pleistocene deposits. Although the Indian subcontinent is nowadays often only thought of as a route towards Java (Hussain et al., 1992), when Asia was believed to be the 'Cradle of Mankind', India and Pakistan were seen as integral to understanding the bigger picture. This is particularly true in reference to the Indonesian fossil remains. Once remains had been discovered on Java, there was an immediate need to explain how animals had arrived on an island many kilometres from the nearest mainland. Various explanations, including sweepstake migrations have been invoked, but the most likely reason may be the existence of the Sunda Shelf resulting in a reduction of the amount of open water to be crossed by colonising animals (see previous section). As a result, considerable attention focussed on the nearby countries of China, India and Pakistan in attempt to resolve the problem of from where the fauna originated. As discussed above, mainland China revealed few sites with fauna of an age comparable to that on Java, but the Siwalik region of India and Pakistan offered a major opportunity to attempt correlations. That is, the Siwalik deposits were studied in the 1920s and 30s because of the Javan remains and not because they were thought to be interesting in their own right.

9.2.1 Background to Siwalik Research

The Siwalik fluvial deposits are named after the Siwalik Hills in Northern India and comprise thousands of metres of sediment deposited along the forefront of the Himalayas and Hindu Kush mountains in the longest and most detailed terrestrial Neogene sequence in the world (Barry and Flynn, 1990; Dennell, 1993; Nanda, 1994). The presence of large numbers of fossils throughout the sequence was documented from the early 19th Century when they were first collected in the Himalayas by Captain Webb and their details published by Dean Buckland. (Buckland is better known in British circles for dismissing the Paviland (male) Upper Palaeolithic burial as a Roman (female) prostitute). The established tradition of fossil collecting was continued by the British military until the mid 19th Century, and the Tertiary age of the fossils was originally established in 1845 by Falconer and Cautley, who also published the first monograph on the Siwaliks. The Geological Survey of India (GSI) was founded in 1860 and several studies of Siwalik material were undertaken under its purview. These resulted in the division of the sequence into two series, the lower of which was unfossiliferous, and correlation of deposits between the Siwalik Hills in India and the Potwar Plateau, now in Pakistan, some several hundred kilometres to the west (Barry, 1995; Dennell, 1993; Dennell, in press).

Further contributions were made by Pilgrim in the early 20th Century when he divided the Siwaliks into three units, Upper, Middle and Lower, and further divided these into various faunal zones (Barry, 1995; Barry and Flynn, 1990; Dennell, 1993; Nanda, 1994). Pilgrim then felt confident in correlating these zones with other faunas across Europe, before writing numerous papers on the fossil taxa recovered. The American Museum of Natural History and Yale and Carnegie Universities were also involved in interpreting the fossil remains found within the Siwaliks and many major syntheses of the material were published before 1939. The result of this intensive research was that by the outbreak of World War II the Siwaliks was the best documented Neogene faunal sequence in the world with fossil collections held in London, New York and Calcutta (Barry, 1995; Dennell, in press and references therein).

Unfortunately, World War II and subsequent partition of India led to a cessation of research until the 1970s. This was mostly due to changing national priorities after Independence, and the fact that although India held a substantial fossil collection, the majority of the type sections upon which the whole chronology of the sequence depended were now located within Pakistan (Dennell, in press; Tandon et al., 1984). Since the 1970s most research into the Siwaliks has been located in

Pakistan where considerable strides have been made in determining the age of the sedimentary sequences through the use of palaeomagnetic and, when possible, radiometric analyses. Prior to this time ageing of the sediments had relied on relative dating and stratigraphical techniques that proved to be inadequate for the task (Barry and Flynn, 1990; Dennell, in press).

9.2.2 Interpreting the Siwalik Sequence – Part 1

The fossiliferous sequence of the Upper Siwaliks consists of fluvial sediments of sandstones, silts and clays deposited in a stream channel (often meandering and braided) that was laterally migrating across a floodplain. Sediment deposition appears to have been cyclical in character, implying periods of basin subsidence and associated fluvial aggregation (Barry and Flynn, 1990; Hussain et al., 1992; Opdyke et al., 1979). Early researchers soon realised that the nature of the deposits meant that the major sedimentary units were discontinuous throughout the Siwaliks, making lithological correlations at both a local and regional level difficult. As a result, they were forced to turn to the fossil remains as a way of dating and relating the various units. Unfortunately, this method was inherently inaccurate, since the faunal remains were frequently highly localised and many areas contained formations with few or no fossils. Nevertheless, little alternative was available to them (Dennell, in press).

Pilgrim divided the Upper Siwalik sequence into various zones on the basis of the faunal remains. Formations that contained no faunal remains were assigned to zones on the basis of lithological similarities with units that contained fossils. This misuse of Pilgrim's faunal zones resulted in formations being assigned to zones to which they did not belong, as well as a general confusion over what the terms (e.g. Nagri, Dhok Pathan, Tatrot etc) actually meant (Barry, 1995; Dennell, in press). It was clear that major improvements in interpretation had to be made before the significance of the faunal remains could be assessed.

9.2.3 Interpreting the Siwalik Sequence – Part 2

Once Siwalik research restarted in the 1970s, two of the key foci were to be able to define better, and to attribute more accurate ages to, the various formations. The fineness of the sequences had enabled researchers to date many accurately through palaeomagnetic analyses. Additionally, although the Siwaliks are mostly comprised of fluvial sediment, volcanic sediments, mostly altered to bentonites and probably originating from the Dash–i–Nawar area of Afghanistan, have been found within some of the sedimentary layers enabling the use of radiometric techniques. This has allowed a programme of fission track analysis of various sequences, chiefly in Pakistan, by both Peshawar and various American Universities. Most of this research has concentrated on the Miocene sequences, particularly between 18 and 7 Ma, found on the Potwar Plateau (Barry, 1995; Dennell, in press; Hussain et al., 1992; Johnson et al., 1981; Johnson et al., 1982a; Johnson et al., 1982b; Opdyke et al., 1979; Tandon et al., 1984).

Geological studies on the eastern part of the Potwar Plateau and around the Jhelum River have shown that long stratigraphic sequences of Upper Siwalik sediments have been preserved. It was thought likely that enough magnetic reversals would have been recorded in the deposits to allow their correlation with the Plio–Pleistocene palaeomagnetic sequence (Opdyke et al., 1979). The presence of volcanic tuffs and bentonites, found at the Upper Tatrot–Lower Pinjor transition, as well as in the earlier Miocene deposits, in the northern Salt Range, aided the interpretation of the palaeomagnetism. In some regions several tuff layers occurred within a relatively short stratigraphic distance allowing the ages of the various sediment layers to be calculated. The ability to use an absolute dating method on the Potwar Plateau was of tremendous importance, since it meant that it was possible to tie the magnetostratigraphic profiles into the known polarity time scale (Johnson et al., 1982a).

It has been shown that the Potwar Siwaliks span over 17 million years with few widespread breaks in deposition, although some local hiatuses are apparent (Barry, 1995; Barry and Flynn, 1990). Magnetostratigraphy has revealed that the vast majority of the Miocene Nagri Formation had normal polarity, a characteristic that made correlating the various magnetic polarity sections relatively easy (Johnson et al., 1982b). It was also possible to date by fission–track some bentonite found within this long interval, and a date of 9.46 ± 0.59 Ma was recorded, firmly placing the Nagri Formation within magnetic chron 9 (Johnson et al., 1982a). A fission–track age obtained from a section in the Pind Savikka and Kotal Kund area has also been important for correlating the magnetic polarity stratigraphy. This section spans a long interval, from the Chinji Formation to the Upper Siwaliks, and so can be linked to the data obtained from the Nagri Formation. A date of 2.5 ± 0.3 Ma was obtained towards the top of the section at a normal–reversed magnetic boundary that has been correlated to the Gauss–Matuyama boundary and is also likely to be associated with the Tatrot/Pinjor faunal boundary (Johnson et al., 1982a; Johnson et al., 1982b).

The use of magnetostratigraphy and fission track has facilitated the construction of a new biostratigraphic framework for the Potwar Plateau, with the field location of fossils being tied into a long, mapped and dated stratigraphic sequence. As a result of this detailed analysis, the ages of fossil sites can be established to within about 200,000 years (Barry and Flynn, 1990) for much of the Miocene, a considerable improvement on the use of the faunal zones. Unfortunately, this level of accuracy has not yet been achieved for the Upper Siwalik formations in Pakistan or the formations in India.

The Geological Survey of India determined that Pilgrim's faunal zones represented faunal stages and so proposed new names for the various lithological formations found within the Upper Siwaliks of India. The Dhamala Formation corresponded to the Tatrot stage, the Kansal Formation to the Pinjor stage and the Tavi Formation to the Lower Boulder Conglomerate (Azzaroli and Napoleone, 1982). Nevertheless, Pilgrim's original terms were still widely used, and often regarded as interchangeable with lithological units (Hussain et al., 1992).

A magnetostratigraphic study was conducted on four sections of the Siwalik sediments in the vicinity of the type locality of Pinjor, near Chandigarh, India. One sample section, at Saketi, covered Pilgrim's Dhok Pathan and Tatrot faunal stages. This sample, including the Dhok Pathan/Tatrot boundary contained several magnetic reversals, but was thought to fall entirely within the Gauss magnetic epoch (Azzaroli and Napoleone, 1982). Further samples of Siwalik sediments were analysed from Masol and Dhamala, and covered the supposed boundary between the Dhamala and Kansal Formations, and thus also represented the transition between the Tatrot and Pinjor faunas in this area. The analysis revealed that all the samples from Masol and Dhamala possessed normal polarity, and so these sections covering the Upper Dhamala and extreme base of the Kansal Formations were assigned to the Gauss magnetic epoch as well. The final section analysed was at Nadah and covered the Pinjor and Boulder Conglomerate faunal stages. The bottom of this section revealed reversed polarity and thus fell within the Matuyama epoch. Sampling did not continue high enough in the section to reveal the transition to the Brunhes chron, but it was apparent that this would occur within the Tavi Formation (Azzaroli and Napoleone, 1982).

These results show, therefore, that in the Pinjor region of the Siwaliks (i.e. in India) the Boulder Conglomerate faunal stage and the vast majority of the Pinjor stage occur within the Matuyama chron, with the Tatrot and Dhok Pathan stages falling within the Gauss chron. The exact transition between the Pinjor and Tatrot stages was unable to be identified within the sections analysed, but was believed to correspond roughly to the switch between the Gauss and Matuyama chrons (Azzaroli and Napoleone, 1982). Additional palaeomagnetic research was carried out in the area east of Chandigarh, but west of Pinjor, to further establish the chronology of deposits within the Siwaliks of this area. A 900 metre thick section of Tatrot and Pinjor beds in the Khetpurali section was analysed. As no volcanic layers were present within the sediments in this area, the magnetic data had to be correlated with biostratigraphic information. This study confirmed the conclusion drawn by Azzaroli and Napoleone (1982) that the Tatrot/Pinjor boundary approximately coincided with the Gauss-Matuyama boundary (Tåndon et al., 1984).

Due to the increasing clarity of the Siwalik fossil record, it was suggested that on the Potwar Plateau Pilgrim's 'faunal zone' terminology should be restricted to lithological units and a new scheme devised to represent the biostratigraphic changes (Barry, 1995; Barry and Flynn, 1990; Nanda, 1994). The base of each of these new zones was characterised by an evolutionary event, specifically the first appearance of an abundant and easily identified immigrant species, and the length of the zone was dependent on the interval before the next evolutionary event. These interval zones were identified in stratigraphic reference sections so that the same zones could be identified in other Siwalik outcrops and also the intervals could be correlated with other phenomena, including magnetostratigraphic studies, allowing an indication of their age. An important feature was that the boundaries of each zone were not fixed, so that if an earlier occurrence of a defining taxon was identified the length of the interval would increase (Barry and Flynn, 1990; Nanda, 1994). However, whilst this method worked very well for the Miocene deposits, the lack of chronological control for later periods meant that there was still much uncertainly over faunal spans after 3 Ma (Dennell, pers. comm. 2001).

Biostratigraphic work was undertaken on the Upper Siwalik sediments of the Mangla–Samwal anticline, Azad Kashmir in Pakistan, believed to date between 3.4 and 1.1 Ma due to correlation with a known palaeomagnetic section (see Johnson et al., 1982a; Opdyke et al., 1979). This work provided a means of understanding the fossil record more fully as a result of the direct relationship between the fossils and known dates and also complemented the work initiated by Barry (1990) (Hussain et al., 1992 and references therein) by attempting to define the more uncertain faunal ranges. It was discovered that Barry's *Elephas planifrons* zone could be extended back in time in this area of the Siwaliks with an interval of 3.3 – 2.7 Ma and a new zone, *Elephas hysudricus* was proposed for the period after 2.7 Ma (Hussain et al., 1992), but with an undefined end point. However, the problem remained that the absence of the indicator fossil did not necessarily signify an evolutionary event, but could reflect the alteration of local vegetation and associated fauna as a result of stream meandering (Hussain et al., 1992).

The above improvements in chronology of the Siwalik deposits have enabled a clearer picture of Miocene palaeogeographical and faunal changes to be developed. Research into the Plio–Pleistocene deposits, however, is still at a more elementary level. Despite fossil remains being tied into local palaeomagnetic sequences, a well-calibrated biostratigraphy is lacking, as discussed below (Dennell, in press).

9.2.4 Plio–Pleistocene Siwalik Deposits

Siwalik sediments less than 7 Ma are generally rather poorer in fossil material than older deposits, and in particular deposits younger than 5 Ma tend not to have been studied as extensively as Miocene deposits. This has meant that much of the Plio-Pleistocene record within the Siwaliks has remained unsampled. Three of Pilgrim's faunal zones fall within this time period, the Tatrot, Pinjor and Boulder Conglomerate Zones, and part of the problem with research in this period relates to his choice of type localities (Barry, 1995; Dennell, in press).

The Tatrot type locality is located on the Potwar Plateau in the Jhelum area of Pakistan and comprises a relatively short section of 60 metres that is both over– and underlain by unconformities. This means that it is extremely difficult to correlate this section with similar sections elsewhere in the Siwaliks, nor is it possible to determine the exact length of time represented by this faunal zone. The fossils that have been found within the Tatrot type section deposit are few in number, further inhibiting regional correlations (Azzaroli and Napoleone, 1982; Dennell, in press). Nevertheless, it has been suggested that the Tatrot Formation could date between 3.2 – 3.4 Ma on the basis of unpublished fossil collections from the Harvard–Geological Survey of Pakistan group (Barry, 1995).

The Pinjor type locality is located c. 400 km southeast of Tatrot near Chandigarh in India (Nanda, 1994), but due to the absence of Pinjor type deposits near Tatrot the exact relationship between the two faunal stages in unclear (Tandon et al., 1984). It would appear, however, that the base of the Pinjor Formation does not correlate with the top of the Tatrot Formation leaving a substantial period of time that is not associated with any geological stage (Dennell, in press). Several fossil localities within the vicinity of the Pinjor type location contain what are believed to be Tatrot–type fauna, but unfortunately older fossil collections make no differentiation between the two faunal stages. It would appear likely that the basal Pinjor Formation contains Tatrot–type fossils, leading to extra complications in understanding the faunal evolution of the area (Azzaroli and Napoleone, 1982; Nanda, 1994). The deposits at Pinjor contain plentiful fossils, as do other locations with sediments attributed to the Pinjor faunal zone. Unfortunately, most fossils lack detailed provenance information and so cannot be correlated with the local palaeomagnetic stratigraphy, resulting in an undifferentiated period of over 2 million years covering the Late Pliocene to Middle Pleistocene (Dennell, in press).

More recent work on the Pinjor Stage in Pakistan has been undertaken in order to understand better the faunal sequences and has resulted in the collection of assemblages from various areas including especially the Mangla–Samwal anticline and the Pabbi Hills, as well as Jalalpur and Rohtas. These new fossils have been linked to the local palaeomagnetic stratigraphy allowing a more detailed appreciation of faunal turnover (Dennell, in press and references therein; Hussain et al., 1992).

The 'Boulder Conglomerate' faunal zone was identified by Pilgrim at several locations within the Siwalik Hills and Salt Range areas and was originally thought to be extremely fossil rich. However, fossils identified as belonging within the Boulder Conglomerate faunal zone were subsequently reassigned by Pilgrim to the Pinjor stage resulting in an almost unfossiliferous faunal zone (Dennell, in press). The most likely explanation for the lack of fossils is that the sediment and associated boulders were deposited under a high–energy regime leading to the destruction of a considerable portion of the animal remains. The age limits of this zone vary widely across the Siwalik region with ages ranging from the Pliocene until the Middle Pleistocene, 3.3 Ma – *c.* 0.4 Ma. As a result of both the lack of fossils and an extremely wide age range, the Boulder Conglomerate faunal zone is not a practicable method of characterising aspects of the Siwalik sequence (Dennell, in press).

9.2.5 Climate Reconstruction

As discussed above, the calibration and quality of information from the Plio–Pleistocene Siwalik formations is of an extremely poor standard and, as a result, palaeoclimatic interpretations based on the faunal and floral remains found within the sediments are highly generalised. Also, an analysis of the faunal remains from north–east of Chandigarh, India revealed that the Upper Siwaliks contained highly biased deposits with assemblages dominated by heavier, denser and more durable skeletal elements. This is because the formation is essentially fluvial in origin, and thus represents an environment where taphonomic effects are especially noticeable (Gaur and Chopra, 1984). Some of the localities from which fossils were collected in the Mangla–Samwal anticline, Pakistan, have also been identified as suffering from the effects of fluvial taphonomic changes (Hussain et al., 1992). This means that the usefulness of fossil remains recovered from these deposits is very limited in terms of climate reconstruction since the fossils are likely to have experienced allochthony (i.e. long-distance transport and sorting by fluvial action), resulting in the elimination of various aspects of the assemblage.

A small collection of Upper Siwalik faunal fossils (minimum number of individuals = 67) has been collected from north–east of Chandigarh and have assisted in interpretations of the palaeoenvironment of the area. A few cold–blooded animals, such as crocodiles and turtles, and some *Hexaprotodon* fossils have been recovered from Tatrot sediments, suggesting a relatively warm environment with much surface water (Gaur and Chopra, 1984). Bovids from this formation are not of typical cursorial or grassland types, but rather show an affinity to bushland, woodland and savanna, a landscape that is confirmed by the presence of species of proboscidean and suid. The general picture of the Tatrot timescale within this area, from this extremely small assemblage, is one of a mosaic environment with wooded grassland, woodlands and some open grassland, but there would appear to be no indication of forest (Gaur and Chopra, 1984).

The Tatrot–Pinjor boundary is marked by a major faunal change in all areas of the Indian Siwaliks. In particular, cervids with antlers make their first appearance in the Pinjor faunal stage, as do *Equus sivalensis*, *Elephas hysudricus*, *Rhinoceros palaeindicus*, *Rhinoceros sivalensis* and *Rucerus simplicidens* (Gaur and Chopra, 1984). The number of crocodilian fossils recovered from the Pinjor stage to the north-east of Chandigarh were even fewer than the preceding Tatrot stage. This could suggest a possible cooling of the climate (Gaur and Chopra, 1984), however, based on fossil collections from Pakistan, is more likely to reflect a change in sampling environment from that of riverine deposits to floodplain (Dennell, pers. comm. 2001), especially when dealing with a meandering braided stream ecosystem. The most commonly found fossils were suids, bovids and elephants, representative of grassland habitats; the most frequently found faunal remains were cursorial animals indicative of open landscapes (Gaur and Chopra, 1984). There is no doubt that the climate experienced considerable change around the Tatrot-Pinjor boundary, which is probably around the Gauss/Matuyama transition, and this has been linked to increasing aridity in the area (Gaur and Chopra, 1984).

Typical Pinjor fossils have also been recovered from Paonta, Himachal Pradesh and include *Equus* and *Bos*, fauna that characterise relatively dry conditions. Whilst Tatrot marker fossils were not recovered from this locality, *Stegodon*, *Archidiskodon* and *Hexaprotodon* were all recovered from beds exhibiting the typical lithology of the Tatrot formation. This fauna would seem to indicate wetter conditions in the preceding Tatrot faunal stage than in the Pinjor stage (Nanda et al., 1991), confirming the evidence from near Chandigarh.

Floral records from the Upper Siwaliks are characteristically scarce and also have poor chronological resolution. Nevertheless, conifer, moss, grass and fern pollen and spores have been recovered from Himachal Pradesh, Ganga Valley, Jammu and Nepalese regions from deposits covering the whole of the Upper Siwaliks, indicating substantial levels of precipitation leading to the development of rivers, swamps and marshes (Mathur, 1984). This is confirmed by the presence of dipterocarps, representative of tropical vegetation, from many different localities within the subcontinent from Miocene until Late Pliocene times. The aridity indicated by the faunal remains would appear to have occurred from the end Pliocene onwards with Dipterocarpaceae becoming restricted to north–eastern and south–western India (Sahni and Mitra, 1980; Singh, 1988).

At present, the best climatic indications are derived from C_3/C_4 analysis of soil carbonates. Quade et al. (1993) shows that the floodplains around Mirpur, Rohtas and the Pabbi Hills were covered by almost pure C_4 grasslands during the Pleistocene, with localised patches of riparian woodlands. Unfortunately, the lack of a sedimentological record for the lowland hills in the region precludes a vegetation reconstruction for that environment. There is, however, evidence for a modern monsoonal climate in the eastern Potwar region at this time (Quade et al., 1993).

9.2.6 Summary

Despite the lack of clarity of the Upper Siwalik deposits, some broad conclusions can be drawn. In general, it would appear that it is difficult to identify climatic change on the basis of faunal and floral remains since the meandering action of the braided rivers would continually alter the local environment so affecting the character of the fauna and associated vegetation. It seems that the faunal change visible at the Tatrot-Pinjor boundary in India is related to climatic and vegetational changes, specifically an increase in aridity leading to the spread of grassland and hence an increase in cursorial and open habitat animals (Barry, 1995). Whilst the exact timing of this transition remains uncertain, a correlation with the Gauss-Matuyama boundary at 2.5 Ma is thought likely and would thus correspond to other patterns of increasing aridity visible within the Old World.

9.3 Conclusions

Plio-Pleistocene deposits in Java and the Siwaliks have several factors in common. Research in both areas commenced relatively early, at a time when Asia was considered to be of prime importance for understanding human evolution. After the Second World War, when Africa was gaining credence as the 'cradle of hominid evolution', research in both areas slowed down, only for interest in the two areas to reemerge in the later 20th Century. The geology of the localities is extremely complicated, leading to immense problems of interpretation, and probably provides a clue as to why the regions are frequently ignored when theories of hominid migration are proposed. Furthermore, the intricacies of the geology and the lack of a detailed chronology result in a poor understanding of climate change, biogeographical and zoological evolution. Nevertheless, important questions regarding early hominids could be settled in this region.

No clear consensus has yet been reached on how the Plio–Pleistocene fauna, including the hominids, were able to reach Java. The favoured interpretation has been a route through India, with a possible land corridor existing across the Sunda Shelf. Claims for early stone tools, perhaps dating to 2 Ma, from Riwat, Pakistan by Dennell and colleagues (Dennell, 1993; Dennell et al., 1988; Rendell et al., 1989; Rendell et al., 1987) could provide evidence to substantiate this hypothesis, but as yet the early dating of the tools has failed to find widespread support. The lack of any other evidence for early hominids on the lands adjacent to Java has meant that this remains one of the key questions for the region.

Chapter 10

Generation of G.I.S. Predictive Model

The previous sections have been important in setting the scene for the development of the model to simulate early hominid colonisation of Eurasia. It is clear from the discussion in Section 3 that a multitude of factors need to be taken into account to ensure the accuracy of the model conclusions, and the steps that were followed are outlined below.

10.1 Equipment and Software

The model was developed on ESRI (Environmental Systems Research Institute, Inc.) ArcView GIS v3.1 using the Spatial Analyst v1.1 and Reproject Grids v4.0 extensions. This programme is capable of displaying both vector and raster datasets which allows greater flexibility of analysis, particularly when developing spatial models, as discussed in Section 2. An added benefit is that this particular programme is 'Windows' compatible, making it more user-friendly and negates the need for special computer systems such as Sun Microstations.

This research also made use of the recently developed Arc-SDM: Spatial Data Modeller software (Kemp et al., 2001). This software was similar in design to the Weights-of-Evidence software that had previously been developed by the same team (Kemp et al., 1999), but also had the facility to use logistic regression, fuzzy logic and neural network analyses (Kemp et al., 2001). The software was specifically designed to operate in conjunction with ArcView GIS and the Spatial Analyst extension and was thus fully compatible with the existing operational requirements.

Originally the Weights-of-Evidence method was combined with GIS technology to allow the mapping of potential mineral sources through the use of both inductive and deductive (expert knowledge) models (Hansen, 2000; Raines et al., 2000). The Weights-of-Evidence technique allows the combination of spatial data from various sources in order to analyse interactions between the dependent and independent variables by examining the statistical relationship between them. In this particular model the weights were calculated from the measured association between known early hominid sites and the predictor environmental variables (Kemp et al., 1999). The procedure is based on Bayes' Rule of Probability, with its implicit assumption of conditional independence (Hansen, 2000; Raines et al., 2000). It can therefore be used to express the probability that a unit of area would contain a hominid site, with a measure of confidence in the results also obtained (Raines et al., 2000).

10.2 Methodology

10.2.1 Time Period and Study Area

As discussed in Section 1, it is commonly believed that early hominids were confined to the African continent for several million years after their evolution until sometime approaching 1 Ma. This study set out to explore the possibility that the exodus could have been geographically possible from an earlier date, closer to 2 Ma. Therefore, faunal, floral, ecological and geographical data were collected from the literature for sites falling within a timespan of about 2.5 – 1.5 Ma, as discussed in previous sections.

In order to compare the geographical situation within Africa to that in the areas to which the hominids could have migrated, it was necessary to collect data from both African and Eurasian localities. So as to achieve a realistic model, data from a vast area had to be collated. Suitable quantities of data were not always easy to obtain, as outlined in Section 4.

10.2.2 Data Layers

Data applicable to this predictive model was obtained from a wide variety of sources.

10.2.2.1 ESRI Data

Accompanying the programme ArcView GIS v3.1 are a variety of ArcView compatible data sources, including geopolitical data of the World. Several of these vector datasets were utilised in order to provide a basemap on which other relevant information could be displayed. These data were world countries in 1998, world lakes, world rivers and a world map background depicting latitude and longitude lines. This information can be found on the ArcView GIS 3.1 installation CD, in the subdirectory 'world', and with the file names cntry98.shp, lakes.shp, rivers.shp, latlong.shp and world30.shp.

The 1998 world countries dataset was provided generalised from ESRI's ArcWorld supplement map data. The world lakes and rivers data was provided generalised from ESRI's ArcWorld 1:3,000,000 and 1:25,000,000 databases. The latitude and longitude data was depicted as a 5° by 5° grid that had been mathematically generated by ESRI using ARC/INFO software and then converted to an ESRI shapefile. The world map background was also mathematically generated by ESRI using ARC/INFO software and then converted to an ESRI shapefile (Environmental Systems Research Institute Inc, 1992 – 1998). Additional data had also been included by ESRI from Oxford University Press, 1993; Rand McNally, 1991 and The Times, 1990.

Excess information was removed from the world countries 1998 data so that it was possible to display only the continental coastal outlines without the individual country borders being depicted. This was to enable the drawing of clearer maps.

10.2.2.2 PRISM2 Data

As was discussed in Section 6, raster datasets developed by the Pliocene Research Investigations and Synoptic Mapping (PRISM2) project were invaluable in indicating the mid-Pliocene conditions of the Northern Hemisphere (Barron et al., 1994; Dowsett et al., 1994; Dowsett et al., 1999; PRISM Project Members, 1996). The project reconstructions included sea levels, land-sea distributions, sea surface temperatures, sea ice and land ice distributions, land surface topography and vegetation distribution for both the mid-Pliocene and modern-day environments (Dowsett et al., 1999). It was decided that for this analysis the only relevant information was land surface topography and vegetation distribution, since this would indirectly provide information about sea-levels and land-ice, whilst sea surface temperature was unlikely to have had an overt effect on continental mammal migration routes. Furthermore, the scarcity of information on these topics would have made it extremely difficult to determine accurately their distributions in the Later Pliocene and Pleistocene.

The PRISM2 data is available online at:
http://chhtntsrv.er.usgs.gov/warmclimates/ofr99_535.html.
The datasets required were Pliocene topography, modern topography, Pliocene vegetation and modern vegetation. The topographic information was recorded in metres above sea level, whilst the vegetation type had been numerically coded, with a value determined for each 2° by 2° segment of the earth, as discussed in Section 6.

Unfortunately, it was not possible to download the data in a format compatible with ArcView GIS v3.1, and so the data was acquired as tab delimited text files. These files were then imported into Microsoft Excel 97, where it was possible to start to edit the data to enable its importation into ArcView.

The files initially consisted of rows and columns of data with each individual point representing a 2° by 2° portion of the earth identified by a co-ordinate pair. Since the PRISM2 database had been developed for the entire earth, the first step in preparing the data was to remove all the information for regions outside the square zone of the research area. This left the data for Africa, Europe and Asia. Next, all the co-ordinate information had to be deleted from the file so that only the data remained. This file was then saved as a text delimited file once again and imported into Microsoft Word 97.

It was then necessary to further prepare the data to allow it to be imported into ArcView GIS in an ASCII raster file format. An ASCII raster file consists of a few lines of header information followed by the list of cell values (data) (Environmental Systems Research Institute Inc, 1992 – 1998), and so although the cell values were now correct, the header information had to be compiled. The header information comprised the number of columns in the dataset (ncols), the number of rows in the dataset (nrows), the x co-ordinate of the lower left-hand cell (xllcorner), the y co-ordinate of the lower left-hand cell (yllcorner), the cell size for the database (cellsize) and the number given to cells whose value was unknown (nodata_value) (Environmental Systems Research Institute Inc, 1992 – 1998). This allows the GIS programme to interpret where each row of data ends. So, for the PRISM2 information the format was as follows: –

```
ncols 105
nrows 63
xllcorner –29
yllcorner –45
cellsize 2
nodata_value –9999
9 7 7 7 7 0 0 0 0 0 0 0 0 0 0 0 0 0 0 0 0 0 7 7 7 7 0 0 0 0 0 0
0 0 0 0 0 0 0 0 0 0 0 0 0 0 0 0 0 0 0 0 0 0 0 0 0 0 0 0 0 0 0 0
0 0 2 0 0 0 0 0 0 0 0 0 0 0 0 0 0 0 0 0 0 0 0 0 0 0 0 0 0 0 0 0
0 0 0 0 0 0 0 0 0 0 0 0 0 0 9 9 7
```

The files were then imported into ArcView GIS v3.1 using the Spatial Analyst v1.1 extension.

As is the case with most forms of electronic data, some errors were visible within the PRISM2 datasets. The only miscalculation that had any impact on this research can be found in the topographic data, where any land lying below sea level was given a value of zero, along with all water bodies. This had the effect of increasing the extent of both the Caspian and Dead Seas and also slightly altered the coastline of the continents. However, this error should not have an effect on the overall conclusions of the model.

The evidence outlined in Sections 7, 8 and 9 has assisted in confirming most of the climatic and geographical conclusions that were drawn by the PRISM2 project. It would appear that the climate of the Later Pliocene was not dramatically different from that of the period 3.29 – 2.97 Ma that was examined by the PRISM2 project (Barron et al., 1994; Dowsett et al., 1994; Dowsett et al., 1999; PRISM Project Members, 1996), and so the PRISM2 vegetation reconstruction was used throughout the model simulation. However, research into the orogenic evolution of the Himalayas has suggested a Late Pliocene height of around half of that found in the region today (see Section 7.1) and so a significant alteration of the PRISM2 Pliocene topography dataset was required. This was achieved by following many of the same steps outlined above. The PRISM2 dataset was downloaded along with the other required datasets and the data pertaining to regions outside the study area was deleted. As the PRISM2 information was collected in 2° by 2° cells, it was necessary to identify which cells were occupied by the Himalayan mountain belt. The following list of co-ordinates identifies the bottom left-hand corner of each 2° by 2° cell found to be mostly covered by the

Himalayas: 37°N, 71°E; 35°N, 71°E; 35°N, 73°E; 35°N, 75°E; 33°N, 75°E; 31°N, 77°E; 31°N, 79°E; 29°N, 81°E; 29°N, 83°E; 27°N, 85°E; 27°N, 87°E; 27°N, 89°E; 27°N, 91°E; 27°N, 93°E; 27°N, 95°E; 29°N, 95°E; 27°N, 97°E. The height recorded by the PRISM2 project in these cells was then divided by two to obtain the revised Plio-Pleistocene elevation of the Himalayas. Once the file had been saved in ASCII format, it was then imported into ArcView GIS v3.1.

10.2.2.3 Data from the Literature

Both floral and faunal data was obtained from the literature, and Table 1 lists the localities from which data was acquired. It was not always possible to identify the exact location from which the data had been collected as few site reports or regional syntheses contained the x,y co-ordinates of the data points. Where possible the relevant co-ordinates were obtained from a variety of atlases or gazetteers. In cases where an accurate location could not be identified co-ordinates were obtained for the centre of the area from which the data could have been collected. Some faunal lists were only available for regions, such as Europe or the Old World, and it was not always clear what sites had been included in the synopsis; so once again co-ordinates were obtained for the centre of the area from which the data could have been collected. These uncertain localities are identified by question marks in Table 1. The problems experienced in this stage of the analysis lend extra weight to the suggestion made by Kidd and Ritchie (2000), that all biogeographical studies should include location co-ordinates as well as names (Kidd and Ritchie, 2000). This would have greatly decreased the time taken to identify the relevant sites and would have been particularly informative in cases where more than one locality shared the same name. The addition of co-ordinate information to site reports should not prove too difficult with the development of GPS technology and would be a step in the right direction with the increasing technological nature of archaeological research.

The floral data included pollen diagrams, floral lists and palaeoenvironmental reconstructions and was invaluable in determining the accuracy of the PRISM2 vegetation reconstructions, particularly in the Saharo-Arabian desert belt. In some instances the same data as that used by the PRISM project was analysed, although in many cases it was also possible to find alternative sources. As was discussed in Sections 7 and 8, the Pliocene vegetation reconstructed by the PRISM2 project (Dowsett et al., 1999) was found to be sufficiently accurate for the Plio-Pleistocene period to be used as a basemap for the hominid migration simulation. No further use of the pollen information was made in the development of the GIS model as accurately assessing the importance of individual pollen genera or species is fraught with difficulty since it is the entire pollen assemblage that characterises vegetation. This topic is discussed in more detail in Section 5.

Faunal data from all investigated sites was used within the GIS simulation as vector point data, and references from which the data was procured are listed in Table 2. Due to the sparse nature of the palaeontological record, it was decided to record all macromammalian remains at the generic level so that faunal correlations between sites could be made more easily. Also, the attribution of species or sub-species names to fragmentary faunal remains can prove to be extremely problematical (See Section 5), and so the use of generic taxonomy was likely to lead to a greater degree of accuracy. A listing of the genera used in the generation of the model can be found in Appendix 1. It was found that many different genera were present at the sites from which faunal remains had been recovered, with little overlap between the sites. That is, most genera were only found at one or two sites, meaning that any analysis undertaken to determine whether particular genera were preferentially associated with hominids would produce meaningless results. Consequently, the data were incorporated into the larger categories of carnivore, carnivore/scavenger, herbivore and herbivore/root/ omnivore diet type, as also listed in Appendix 1. Therefore, for each site the total number of genera present, divided into different diet type groups, was known.

Table 1. Geographical Location of All Sites From Which Faunal and/or Floral Data were Obtained from the Literature.

Site	Country	Latitude (N)	Longitude (E)
Africa	Africa	? 5.00	? 20.00
Ahl Al Oughlam Quarry	Morocco	33.62	-7.58
An Nafud	Saudi Arabia	27.75	39.50
Apollonia	Greece	40.67	23.25
Aramis	Ethiopia	9.50	41.00
Atapuerca - TD6	Spain	42.50	-3.50
Bahr el Ghazal	Chad	15.96	18.86
Betfia 2, Crisana	Romania	47.00	22.00
Bethlehem	Levant	31.67	35.25
Bijiashan	China	25.68	109.18
Bouri Peninsular	Ethiopia	10.25	40.56
Carpathian Basin	Hungary	47.00	20.00
Central Europe	Central Europe	? 48.00	? 15.00
Chambal Ridge	Pakistan	32.70	73.42
Chandigarh	India	30.83	76.83
Chemeron Formation	Kenya	0.50	36.00
Chiaijiashan	China	39.75	118.16
Dachai	China	37.50	112.50
Daferno	Greece	40.16	21.67
Damiao	South China	31.00	109.00
Deutsch - Altenburg 2	Austria	48.00	16.50
Dmanisi	Georgia	44.33	41.33
Dongpaoshan	China	? 28.00	? 100.00
Dongyaozitou	China	39.75	115.00
Drama Basin	Greece	41.10	24.10
El - Ghab, Western Flank Depression	Syria	35.50	36.30
Elis	Greece	37.70	21.30
Europe	Europe	? 48.00	? 15.00
Fejej	Ethiopia	4.50	36.00
Fuenta Nueva - 3, Orce	Spain	37.67	-2.33
Ganga Valley	Himalayan Foothills	30.00	78.20
Gaopen	China	30.00	111.00
Gaoping	China	31.00	110.00
Gerakarou	Greece	40.67	23.25
Gesher Benot Ya'aqov	Levant	33.00	35.00
Gonghe	China	36.00	100.00
Grotte Zabia	Poland	51.00	19.50
Hadar	Ethiopia	10.08	42.13
Hemugou	China	? 45.00	? 102.00
Himachal Pradesh	Himalayan Foothills	31.83	77.00
Hohwa Member	Uganda	2.00	32.00
Holstein	Czech Republic	49.67	17.00
Houhecun	China	34.83	110.00
Huangkan	China	40.00	117.00
Huiyu	China	39.00	116.00
Huoshan	China	31.50	116.33

Table 1 cont. Geographical Location of All Sites From Which Faunal and/or Floral Data were Obtained from the Literature.

Site	Country	Latitude (N)	Longitude (E)
Ioannina	Greece	39.70	20.70
Jammu - Kutwatta - Chiani Section	Himalayan Foothills	32.83	75.00
Jiajiashan	China	? 45.00	? 102.00
Jing - Le Basin	China	38.00	113.50
Jingou	China	? 45.00	? 102.00
Jisr - Ash - Shaghour	Syria	35.75	36.25
Jockgrim	Germany	49.00	8.33
Juyuangong	China	? 28.00	? 106.00
Kamyk	Poland	50.83	19.00
Kanjera Formation	Kenya	-0.33	34.50
Kielniki 3A	Poland	52.50	19.75
Kleszczów Graben	Poland	51.07	19.07
Kokiselei	Kenya	4.00	35.91
Koobi Fora	Kenya	4.50	36.50
Krimni - 1	Greece	40.67	23.25
Kurkur Oasis	Libya	23.90	32.32
Laetoli	Tanzania	3.02	34.15
Lantian	China	31.00	110.00
Latamne	Syria	35.00	37.00
Leffe Basin	Italy	45.82	9.85
Lengzhaishan	China	? 24.00	? 109.00
Lokalalei	Kenya	4.00	35.91
Longgupo	China	30.75	108.67
Lower Karewa	India	30.00	76.00
Mangla - Samwal Anticline	Pakistan	33.10	73.77
Meiningen	Germany	50.67	10.42
Melekh Sdom	Israel	31.00	35.00
Middle Awash Valley	Ethiopia	9.00	40.50
Mosbach 1	Germany	49.42	9.08
Nagharsányhegy 2	Hungary	45.67	18.50
Naiyena Engol	Kenya	4.00	35.91
Neapolis	Greece	40.20	21.00
North Africa & Middle East	N. Africa & Middle East	? 25.00	? 25.00
North Asia	North Asia	? 45.00	? 83.00
North China	China	? 45.00	? 102.00
North China Region	China	? 45.00	? 102.00
North of Qinling Mountains	China	? 39.00	? 87.00
ODP Site 658	Africa	20.75	-18.58
Old World	Old World	? 48.00	? 58.00
Olduvai Gorge	Tanzania	-2.15	34.12
Omo	Ethiopia	5.00	36.07
Osztramos 2	Hungary	? 47.00	? 19.50
Pabbi Hills	Pakistan	32.80	73.80
Paonta	India	29.00	77.58
Potou	China	? 45.00	? 102.00
Ptolemais	Greece	40.60	21.20

Table 1 cont. Geographical Location of All Sites From Which Faunal and/or Floral Data were Obtained from the Literature.

Site	Country	Latitude (N)	Longitude (E)
Pula Maar	Hungary	46.83	17.50
Pyrgos	Greece	37.70	21.30
Ravin of Voulgarakis	Greece	40.67	23.25
Rohtas Anticline	Pakistan	32.90	73.58
Saint - Macaire Maar	S. France	43.40	13.32
Saropée	Thailand	15.50	102.83
Semaforo Section	Italy	39.02	16.70
Semliki Valley	Zaire	0.00	30.00
Serres	Greece	41.00	23.80
Shijiawan Section	China	34.50	109.75
Site 642, Punjab.	Pakistan	32.00	73.00
Slivnitsa	Bulgaria	43.00	23.00
South Asia	South Asia	? 29.00	? 99.00
South China	China	? 28.00	? 106.00
South China Region	China	? 28.00	? 106.00
Southern Sicily & Southern Calabria	Italy	? 37.50	? 14.50
Sub - Saharan Africa	Africa	? -4.00	? 26.50
Tatrot	Pakistan	32.83	73.35
Tharukhola - Chepang Section	Nepal	? 28.00	? 84.50
Tianqiao	China	? 28.00	? 106.00
Transitional Region, China	China	? 30.00	? 115.00
Tsotyli	Greece	40.20	21.00
Turkana	Kenya	4.00	36.00
'Ubeidiya	Israel	32.75	35.00
Upper Siwaliks	Pakistan	32.82	73.77
Uraha	Malawi	-10.08	34.00
Vassiloudi	Greece	40.67	23.25
Villány 5	Hungary	45.85	18.43
Vlachiotis	Greece	36.80	22.80
Volax	Greece	41.20	24.00
W. Asia & Europe	W. Asia & Europe	? 50.00	? 18.75
Xiashagou	China	40.00	114.00
Yinshan	China	24.50	108.67
Yuanmou Formation	China	25.67	101.83
Zoukoudian	China	37.00	116.00

In cases where an accurate location could not be identified, or where the data covered a wide region, co-ordinates were obtained for the centre of the area from which the data could have been collected.
'?' indicated that the exact location of the site is unknown. The co-ordinates represent the centre point of the area to which the data refers.

Table 2. References for All Sites from which Faunal Data was Procured From the Literature.

Site	References
Africa	Foley, 1984
Ahl Al Oughlam Quarry	Raynal et al., 1990
An Nafud	Thomas et al., 1998
Apollonia	Kostopoulos and Koufos, 1994
Aramis	WoldeGabriel et al., 1994
Atapuerca - TD6	Carbonell et al., 1995
Bahr el Ghazal	Brunet et al., 1995
Betfia 2, Crisana	Wolsan, 1993
Bethlehem	Bar - Yosef, 1994
Bijiashan	Zeng Shaohua and Han Defen, 1991
Bouri Peninsular	de Heinzelin et al., 1999
Central Europe	Wolsan, 1993
Chambal Ridge	Opdyke et al., 1979
Chandigarh	Gaur and Chopra, 1984
Chemeron Formation	Hill et al., 1992
Chiaijiashan	Wanpo, 1991
Dachai	Zeng Shaohua and Han Defen, 1991
Daferno	Kostopoulos and Koufos, 1994
Damiao	Wanpo, 1991
Deutsch - Altenburg 2	Wolsan, 1993
Dmanisi	Bar - Yosef, 1995; Gabunia and Vekua, 1995; Vekua, 1995
Dongpaoshan	Zeng Shaohua and Han Defen, 1991
Dongyaozitou	Zeng Shaohua and Han Defen, 1991
Drama Basin	Tsoukala, 1992 and references therein
Elis	Tsoukala, 1992 and references therein
Europe	Bar - Yosef, 1994; Turner, 1992
Fejej	Asfaw et al., 1991
Fuenta Nueva - 3, Orce	Martínez Navarro et al., 1997
Gaopen	Wanpo, 1991
Gaoping	Zeng Shaohua and Han Defen, 1991
Gerakarou	Kostopoulos and Koufos, 1994
Gesher Benot Ya'aqov	Bar - Yosef, 1995
Gonghe	Zeng Shaohua and Han Defen, 1991
Grotte Zabia	Wolsan, 1993
Hadar	Kalb et al., 1982
Hemugou	Zeng Shaohua and Han Defen, 1991
Himachal Pradesh	Badam, 1984
Holstein	Wolsan, 1993
Houhecun	Zeng Shaohua and Han Defen, 1991
Huangkan	Zeng Shaohua and Han Defen, 1991
Huiyu	Wanpo, 1991
Ioannina	Tsoukala, 1992 and references therein
Jiajiashan	Zeng Shaohua and Han Defen, 1991
Jing - Le Basin	Duowen and Derbyshire, 1991
Jingou	Zeng Shaohua and Han Defen, 1991
Jockgrim	Wolsan, 1993
Juyuangong	Zeng Shaohua and Han Defen, 1991
Kamyk	Wolsan, 1993
Kanjera Formation	Ditchfield et al., 1999; Plummer et al., 1999

Table 2 cont. References for All Sites from which Faunal Data was Procured From the Literature.

Site	References
Kielniki 3A	Wolsan, 1993
Kokiselei	Kibunjia et al., 1992
Koobi Fora	Beden, 1983; Day, 1976; Eisenmann, 1983; Harris et al., 1980; Harris, 1978; Harris, 1983a, b, c; Isaac and Harris, 1978; Sept, 1986; Lewis, 1997
Krimni - 1	Kostopoulos and Koufos, 1994
Lantian	Wu Rukang and Xinzhi, 1984
Latamne	Bar - Yosef, 1994
Lengzhaishan	Wanpo, 1991
Lokalalei	Kibunjia et al., 1992
Longgupo	Zeng Shaohua and Han Defen, 1991
Lower Karewa	Badam, 1984
Mangla - Samwal Anticline	Hussain et al., 1992; Opdyke et al., 1979
Meiningen	Wolsan, 1993
Mosbach 1	Wolsan, 1993
Nagharsányhegy 2	Wolsan, 1993
Naiyena Engol	Kibunjia et al., 1992
Neapolis	Tsoukala, 1992 and references therein
North Africa & Middle East	Maglio, 1975
North Asia	Maglio, 1975
North China	Zeng Shaohua and Han Defen, 1991
North China Region	Xue Xiangxi and Zhang Yunxiang, 1991
North of Qinling Mountains	Guo Shengqiao and Shao Shixiong, 1991
Old World	Maglio, 1975
Olduvai Gorge	Hay, 1971, 1994; Leakey, 1971a, b, c; Leakey, 1994a, b; Lewis, 1997; Monahan, 1996; Tobias, 1991
Omo	Coppens and Howell, 1976; Feibel et al., 1989; Howell et al., 1987
Osztramos 2	Wolsan, 1993
Pabbi Hills	Opdyke et al., 1979
Potou	Zeng Shaohua and Han Defen, 1991
Ptolemais	Tsoukala, 1992 and references therein
Pyrgos	Tsoukala, 1992 and references therein
Ravin of Voulgarakis	Kostopoulos and Koufos, 1994
Rohtas Anticline	Opdyke et al., 1979
Semliki Valley	Boaz et al., 1992
Serres	Tsoukala, 1992 and references therein
Site 642, Punjab.	Ashfaque and Saleem-ul-Haq, 1991; Dennell, in press
Slivnitsa	Spassov, 1998
South Asia	Maglio, 1975
South China	Zeng Shaohua and Han Defen, 1991
South China Region	Xue Xiangxi and Zhang Yunxiang, 1991
Sub - Saharan Africa	Maglio, 1975
Tatrot	Opdyke et al., 1979
Tianqiao	Zeng Shaohua and Han Defen, 1991
Transitional Region, China	Xue Xiangxi and Zhang Yunxiang, 1991
Tsotyli	Tsoukala, 1992 and references therein
'Ubeidiya	Bar - Yosef, 1995; Tchernov, 1987, 1988
Upper Siwaliks	Turner, in prep
Uraha	Bromage et al., 1995a, b; Kullmer et al., 1999; Schrenk et al., 1993
Vassiloudi	Kostopoulos and Koufos, 1994
Villány 5	Wolsan, 1993

Table 2 cont. References for All Sites from which Faunal Data was Procured From the Literature.

Site	References
Vlachiotis	Tsoukala, 1992 and references therein
Volax	Tsoukala, 1992 and references therein
W. Asia & Europe	Maglio, 1975
Xiashagou	Zeng Shaohua and Han Defen, 1991
Yinshan	Zeng Shaohua and Han Defen, 1991
Yuanmou Formation	Qian Fang, 1991
Zoukoudian	Zeng Shaohua and Han Defen, 1991

10.2.3 Projection of G.I.S. Data

In order to analyse the data gathered within the GIS environment it was necessary to project the PRISM2 vegetation and topographic data, since the original data was mapped using geographical co-ordinates and so was not projected. A map projection is an arrangement of meridians, lines of longitude, and parallels, lines of latitude, used to portray the curved surface of the earth (Maling, 1992). ArcView GIS v3.1 has several inbuilt projections from which to choose, and the area of the earth to be mapped and the questions that the map will be used to answer determine what projection would be the most appropriate.

There is no direct correspondence between a place on the earth and a point on a map as some error is always involved in the mapping procedure. Firstly, a scale change has to occur, since it is impossible to depict the earth on a full-scale map, and secondly, to fit the curved surface of the earth onto a plane (i.e. paper) some distortion is inevitable (Maling, 1992). A smaller area of the earth can be depicted more accurately than a larger one as to all intents and purposes it is flat, that is the curvature of the earth does not have to be accounted for. However, as in this analysis, when a large section of the world is displayed on a small-scale map it is necessary to choose a projection that limits the amount of distortion (Maling, 1992).

Distortion occurs in two directions, either through exaggeration of area or angular deformation. Equal-area, or equivalent, and conformal projections are opposite ends of the spectrum in terms of types of projection and both types suffer from large amounts of distortion, but in different areas. Equal-area projections, as the name suggests, are the best way of maintaining the accurate area of a region and are used for statistical distribution maps as well as in analyses where calculating areas is fundamental. Conformal projections maintain angular measurements, but whilst the shape of an object is preserved the size is altered. As a result this type of projection is most suitable for measuring distances and is used in navigational and topographical maps (Maling, 1992). Therefore, when developing the predictive model for hominid site location it is apparent that the projection used should be of an equal-area type, since a statistical distribution map will be produced.

An accurate scale can only be maintained at certain points or along particular lines of the map and these places are known as lines of zero distortion or principal scale (Maling, 1992). The fundamental principle behind projections and associated distortions is that the amount of distortion increases from the origin of the projection to the edge of the map. This means that the area located at the origin of the projection is more accurately represented than those areas further away and so the most important area in terms of the analysis should be located at the centre of the map. The origin is known as the point of zero distortion because no distortion is apparent. With a line or lines of zero distortion the amount of misrepresentation increases outwards from the line to the edge of the map (Maling, 1992). If two lines of distortion or standard parallels are used the amount of distortion is reduced since the area between the lines is portrayed accurately although, as before, the amount of distortion increases outwards from the lines to the edges of the map (Environmental Systems Research Institute Inc, 1992 – 1998). The three different classes of projection take advantage of this effect with a normal aspect cylindrical projection having the line of zero distortion on the equator, a normal aspect conical projection having the line of zero distortion along a line of latitude and a normal aspect azimuthal projection having a point of zero distortion centred on a geographical pole (Maling, 1992).

The shape of the area to be mapped affects the choice of class of projection. In this analysis the land area under consideration comprised the East African Rift Valleys, North Africa, the circum-Mediterranean area, the Levant, Southern Europe and a wide band across the Asian continent between 15° and 40° north. The long axis of this area runs east - west along a line of latitude and so the line of zero distortion should also follow this pattern, indicating that the best class of projection for this project would be conical. A cylindrical projection would not be appropriate in this instance as the majority of the area to be mapped falls outside the tropics (Maling, 1992).

Having shown that the best map projection to use when analysing the data within ArcView GIS v3.1 would be of a conical equal-area type, it was necessary to determine which of the various projections available within the GIS fulfilled the criteria. The Alber's Equal-Area map projection seemed to be most appropriate for this analysis. The projection was described as 'being most useful for mapping an area that extends furthest longitudinally rather than north-south, all regions depicted are proportionally the same area as on the earth and north-south latitudinal extent should not exceed 35°' (Environmental Systems Research Institute Inc, 1992 – 1998). The main area of interest in this research fell between 5° – 40° north, and so complied with this requirement. Although some sites have been recorded at higher and lower latitudes, they were of secondary concern in developing the site predictive model. This was because hominids were unlikely to migrate to latitudes higher than 40° during the time period under consideration due to environmental hazards, and the early presence of hominids close to the equator in Africa has been well documented in the past.

As was described above, it was necessary to centre the projection on the mid-point of the area to be mapped; in this case the central meridian fell at 60° E and the reference latitude at 23° N. The projection utilised two standard parallels or lines of zero distortion. There was more than one way to calculate where these parallels should be drawn, although, technically, the parallels should fall halfway between the origin of the projection and the edge of the map (Maling, 1992). This placed the first standard parallel at 13.75° N and the second at 31.25° N. As the map was to be drawn at a small-scale, it was not necessary to calculate the projection using formulae that define the earth as a spheroid. At the required resolution the difference between a sphere and a spheroid was negligible and so the simpler calculation could be used (Environmental Systems Research Institute Inc, 1992 – 1998).

The projection of the PRISM2 data was achieved by setting the appropriate projection requirements for Alber's Equal-Area projection, as described above, and then running the Reproject Grids v4.0 extension. The reproject function was achieved through the use of polynomial equations that approximate the reprojection. A low transformation order of 1 was employed in order to limit unnatural curving of the map. As the PRISM2 data comprised categorical values, the most appropriate interpolation method was 'nearest neighbour'. Lastly, the same cell size was used on both maps. However, as a result of the polynomial approximation, accuracy was compromised along the edges of the map (Huber, 1999). This error was particularly noticeable when plotting site co-ordinates on both the projected and unprojected maps, as the site locations appeared to move. Fortunately, as the area of most interest to this research fell towards the centre of the map, this did not pose too much of a drawback. Once an appropriate projection for the model had been set, it was possible to start importing the relevant data.

10.3 Arc-SDM: Spatial Data Modeller

The Arc-SDM software was used in order to investigate the hypothesis that certain areas outside Africa would have represented favourable locations for Plio-Pleistocene hominids to migrate into. In particular, these locations could be identified on the basis of vegetation, topography and the characteristics of the macromammalian genera present.

10.3.1 Preparing the Data

As was described above, the appropriate raster ESRI datasets were imported to provide a basemap of continental outlines on which the model results could be displayed. The datasets integral to the model development were also imported, and these were the raster PRISM2 datasets of Pliocene topography and vegetation, point vector files of localities from which carnivore, carnivore/scavenger, herbivore and herbivore/root/omnivore genera had been found and point vector files of hominid sites.

The next step in preparing the data was to convert the faunal point vector files into raster grid files that would show an estimate of the distribution of the different faunal types over the entire research area; this was achieved by utilising the interpolation command within ArcView GIS spatial analyst. Two interpolation methods were available within the ArcView GIS interface, spline and IDW (inverse distance weighted), and the suitability of each technique was dependent upon the type of data to be extrapolated. When using the interpolation technique it was not possible to have more than one data source at any location, and so for sites such as Olduvai Gorge, where the numbers of faunal specimens falling into each diet type category were known for different time periods, the numbers were averaged. In a number of cases, where the exact location of a site was unknown, several sites had been attributed to the same geographical location. These co-ordinates were then altered by 0.05° which, as a result of the scale at which the model was to be generated, would have no overall affect on the model results.

It was determined that the IDW method would be unsuitable in this instance since it was not known if the data points represented the fullest possible range of values and the number of data points was extremely small, only 105 (see Section 4 for an understanding of the paucity of data). The spline interpolation method is known for achieving better results with a small dataset and is appropriate for low variability surfaces, so this was the method chosen. Despite not knowing the full extent of genera numbers for each diet type that could be expected at a site, it was felt that a reasonable range was represented and so the tension spline technique was utilised as the results obtained mimic closely the original sample (Environmental Systems Research Institute Inc, 1992 – 1998). The four vector files containing information regarding diet type were then all interpolated using the tension spline technique, where all data points were included in each calculation and a weight of tension of 0.1 was used. The continental outline layer was used as a mask for this analysis. This had the effect of ensuring that the extent of each interpolated surface was the same as that of the Pliocene vegetation and topographical layers and that areas covered by water would not return a value for faunal genera. In all instances of interpolation, values returned for areas towards the edges of the extrapolated surface were less reliable than those obtained for the centre (Environmental Systems Research Institute Inc, 1992 – 1998). However, in this case the drawback of this method was not relevant since most of the research area fell in the middle of the interpolated surface.

One final step was necessary before the faunal data layers could be incorporated into the GIS model. The interpolation technique produced raster grids of floating data type, that is the grids were comprised of numbers containing decimal places. Since the theoretical number of alternatives was infinite, no data or attribute file containing the actual results was produced and, unfortunately, this attribute file was critical to the functioning of the Arc-SDM model. This problem can be circumvented by first exporting the ASCII data file and then importing it, but asking the software to convert each number to an integer data type (i.e. to undertake 'rounding'). In this way an attribute file was created and the loss of data was negligible.

Illustrations 8 – 11 show the interpolated surfaces for all four of the diet type categories. It can be seen that all four of the illustrations include negative values and, obviously, it is not possible for a location to have a negative number of genera present. However, these negative values were found in the same places on each of the four figures; that is in Northern Russia and Siberia, the heart of the Saharan Desert, Western Asia including Iraq and Iran and a small section of Eastern China. A couple of factors link all of these places. Firstly, some were towards the edge of the interpolated area where it was known that accurate results were difficult to achieve and others were some distance away from the sample locations from which the interpolation surface was generated. Undoubtedly, the availability of more data locations would have increased the accuracy of the extrapolated surface. It would also have been possible to prevent negative values being returned by utilising the IDW interpolation technique, however, as discussed above, the lack of certainty regarding the upper values rendered this method unsuitable. Furthermore, as will be explained in more detail below, the Arc-SDM model required the reclassification of data themes and so the actual values returned were of less importance than the overall impression of areas of high, medium and low density of genera of different diet types.

Illustration 8. Interpolated Map of Herbivore Diet Type Category.

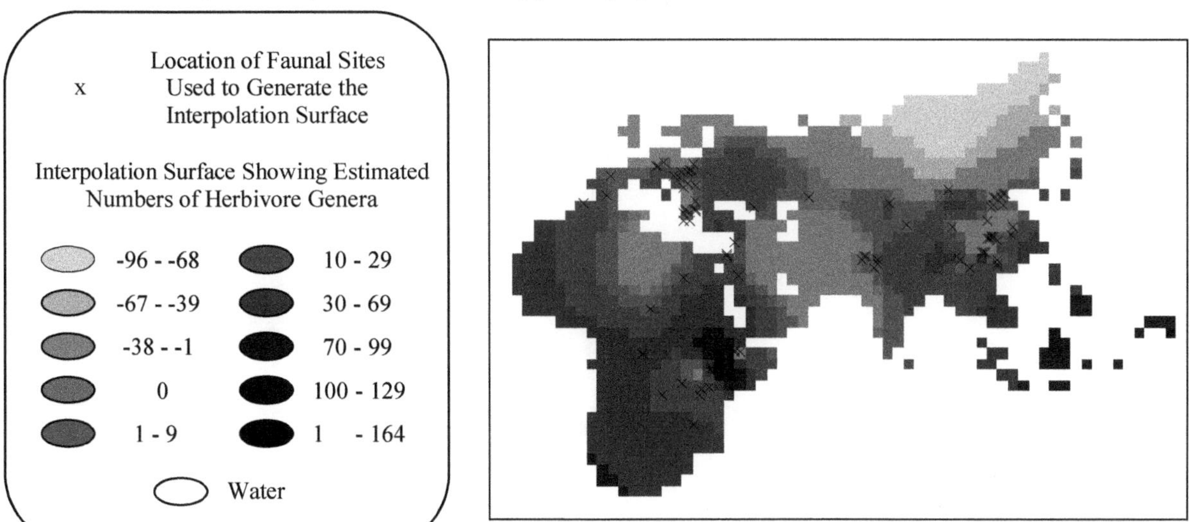

Illustration 9. Interpolated Map of Herbivore/Root/Omnivore Diet Type Category.

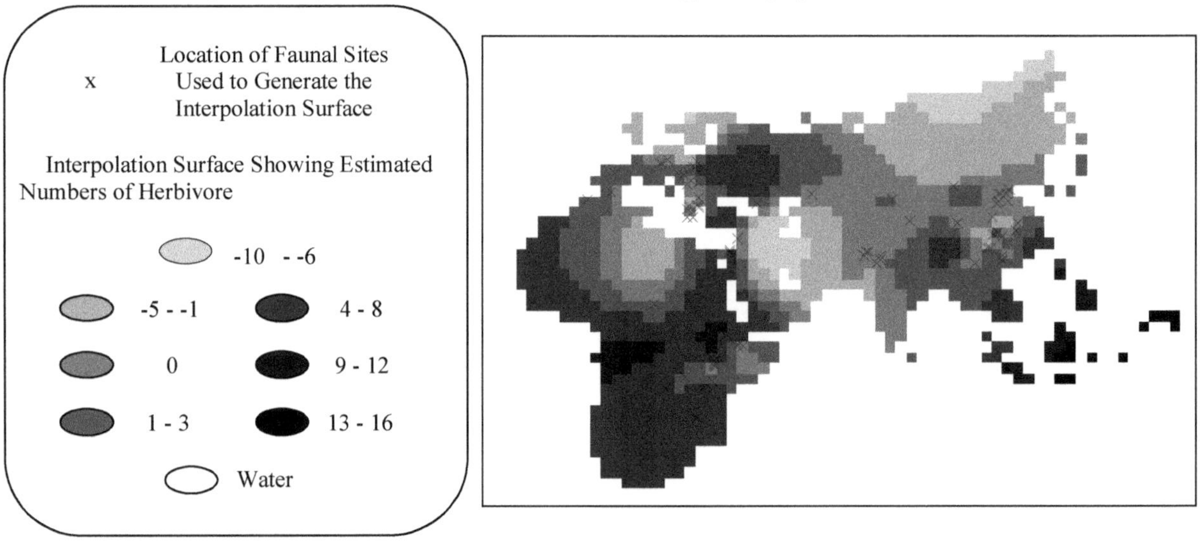

Illustration 10. Interpolated Map of Carnivore Diet Type Category.

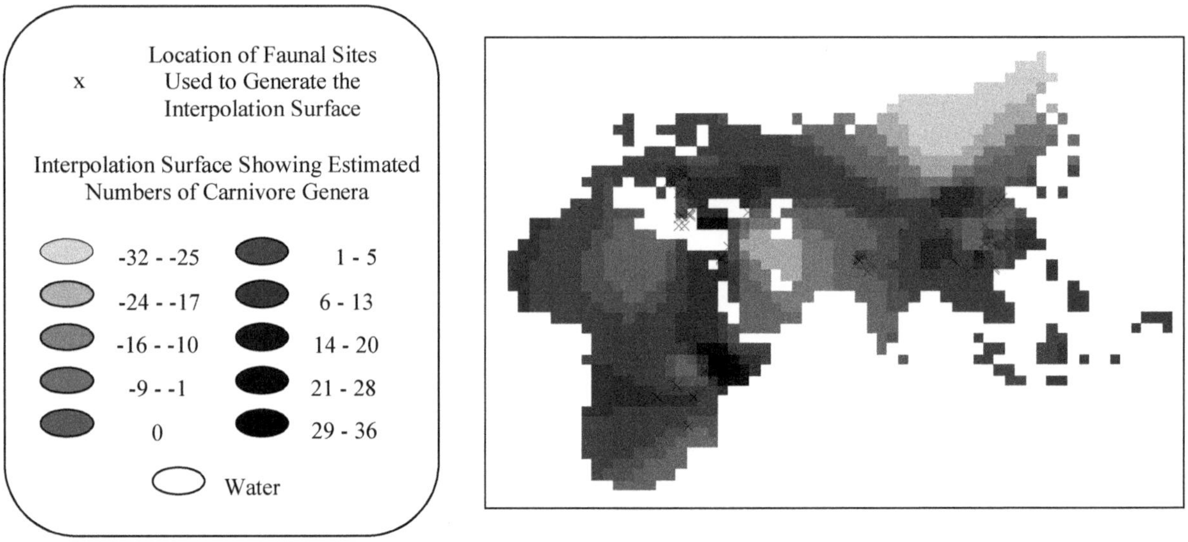

Illustration 11. Interpolated Map of Carnivore/Scavenger Diet Type Category.

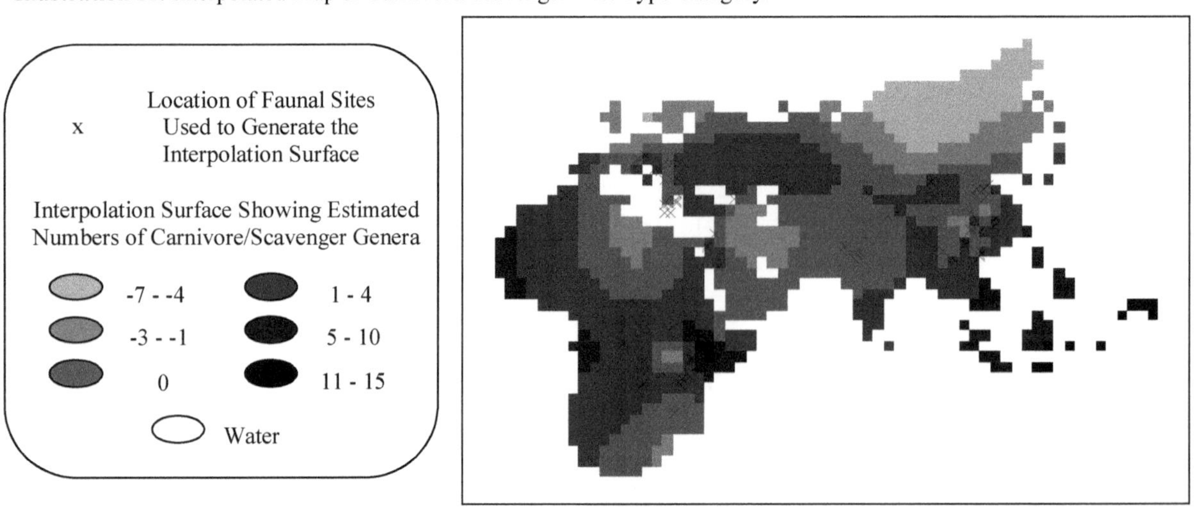

10.3.2 Weights-of-Evidence

The Arc-SDM programme includes four different types of quantitative analysis; Weights-of-Evidence, logistic regression, neural network and fuzzy logic. In this particular instance the only appropriate tool was Weights-of-Evidence, since its application does not require the identification of non-site locations. The current limited knowledge of factors affecting early hominid site preference make it impossible to identify areas where hominids could definitely not have been present, and so using this characteristic in the generation of the model would have rendered the results erroneous.

The process of generating the model can be divided into several steps, not all of which need to be undertaken within the Weights-of-Evidence environment. Firstly, a spatial digital database needs to be built, then the predictive evidence for a particular theme can be extracted, based on an exploration model, by calculating the weights for each predictive map or evidential theme. Finally the evidential themes are combined to predict the potential of locations for the presence of early hominids (Kemp et al., 1999). 'In the final stage, maps can be produced showing a measure of confidence in the results, based on the uncertainty due to the variances of weights' (Kemp et al., 1999).

For this model the spatial digital database was built as outlined above and so the predictive evidence then needed to be extracted. This was achieved by first defining the parameters of the analysis. A mask, demarcating the area to which the model would be applied was identified; this was the ESRI continental outlines theme that recognised areas of water as being outside the study area. Next the entire research area had to be divided into individual unit areas so that the presence or absence of the dependent and independent variables could be calculated for each unit. Due to the coarse nature of the data obtained, the unit area was defined as cells of 15,000 km^2. Lastly, the vector point theme containing the training points, or hominid site localities, had to be named. The hominid sites used to generate the model are listed in Table 3. Only two sites outside Africa, Dmanisi and 'Ubeidiya, were used since it was felt that at this stage the evidence for Plio-Pleistocene hominids at other sites in Eurasia was too equivocal for a model to be based on them. Not all known Plio-Pleistocene hominid sites in East Africa were included in the model generation either. This was due to the large scale at which the model was operating; if too many sites in a region were included in the model then problems linked to ecological dependency would have been encountered. Finally, hominid sites from South Africa were also excluded from the model so that they could be used to test the validity of the results. The weights for every value within the six themes could then be calculated, and the results are shown in Appendix 2.

The calculated weights indicate the amount of association between the training points – hominid sites – and the evidential themes – vegetation, topography and diet type (Kemp et al., 1999). 'The estimate of the prior probability of the occurrence of a hominid site is based on the total number of identified sites, in this case 17, distributed over the entire research area. The posterior probability for each class within the theme is based on the prior probability and the co-occurrence of the class with a hominid site. The odds of this occurring are reflected in the weight value' (Hansen, 2000: 4). The importance of this technique is that it takes the area of each class into consideration; so if a particular class (value) occupied

Table 3. Geographical Location of All Hominid Sites From Which the Arc-SDM Model was Generated.

Site Code	Site	Country	Latitude (N)	Longitude (E)
4	Aramis	Ethiopia	9.50	41.00
6	Bahr el Ghazal	Chad	15.96	18.86
10	Bouri Peninsular, S. Afar	Ethiopia	10.25	40.56
15	Chemeron Formation	Kenya	0.50	36.00
21	Dmanisi	Georgia	44.33	41.33
26	Fejej	Ethiopia	4.50	36.00
35a	Hadar. Fauna 2.5 - 2.0 Ma	Ethiopia	10.08	42.13
51	Kanjera Formation	Kenya	-0.33	34.50
55	Koobi Fora	Kenya	4.50	36.50
62	Lokalalei	Kenya	4.10	35.91
71	Naiyena Engol	Kenya	4.05	35.91
80	Olduvai Gorge. 2.5 - 1.6 Ma	Tanzania	-2.15	34.12
81	Omo. 3.0 - 1.0 Ma	Ethiopia	5.00	36.07
92	Semliki Valley, Lusso Beds	Zaire	0.00	30.00
98	Sub - Saharan Africa	Sub - Saharan Africa	-4.00	26.50
103	'Ubeidiya	Israel	31.99	35.00
104	Uraha, Chiwondo Beds	Malawi	-10.08	34.00

more than 50% of the area it would be expected that more than 50% of the training points would be found in that class. 'Therefore, a positive weight value reflects that there were more points present on that class than would be expected due to chance and vice versa. A value of zero, or very close to zero, indicates that the training points were randomly distributed' (Kemp et al., 1999). 'In general, weights between 0.1 and 0.5 are mildly predictive; values between 0.5 and 1 are moderately predictive; values between 1 and 2 are strongly predictive, and weights greater than 2 are extremely predictive' (Hansen, 2000: 7).

Appendix 2 shows that for all diet type categories except herbivore, values below and including zero were not predictive when determining hominid site location. For carnivore and carnivore/scavenger genera there was an overall trend for the predictive value of the weight to increase as the number of genera increased. The situation was not as clear-cut with herbivorous genera. High and low numbers of genera, including some negative numbers, were predictive when locating hominid sites, with an increasing number of genera resulting in an increased weight, but a mid-range number of herbivore genera, between 18 and 51, was not predictive. In general herbivore/root/omnivore genera were not useful when predicting hominid site location. Only two types of vegetation, grassland and coniferous forest, were useful when detecting early hominid sites. The topographic variable was more complicated to interpret. It appears that hominid sites were preferentially located in areas of moderate height, as land between 900m and 1280m had a high correlation with site locality. Land above *c.* 1550m had no predictive value in determining site location. On the basis of this information the six independent variable themes were reclassified or generalised, following a binary pattern, into predictive and non - predictive values as listed in Appendix 2.

The data from the six evidential themes was then ready to be combined into a response theme, utilising a Bayesian approach, in order to predict areas of potential hominid migration within the research area.

10.3.3 Results

As well as a graphical display of the resulting response theme illustrating the posterior probability surface (Hansen, 2000), shown in Illustration 12, the Arc-SDM software package also returned various tables detailing the Bayesian statistical underpinnings of the display. The response theme classified unit areas that met the characteristics identified and defined by the weights calculated in the exploration model; that is areas that had the same features as the hominid sites used to generate the model.

In order to evaluate the model results it was necessary to show that the premise on which the model was generated was accurate. That is that the independent themes on which the model was based were truly conditionally independent of each other in respect of the hominid sites (Bonham-Carter et al., 1989). Tables 4 and 5 show the results of a Chi-squared (χ^2) test for each combination of the binary themes and a probability table to assess the level of conditional independence. Only one pairing, that of Herbivore/Root/Omnivore and Carnivore diet types, returned a significant χ^2 response along with a probability of zero, indicating a problem with conditional independence. As it was likely that this response was due to an extremely low expected frequency, related to the small number of training points used in the generation of the model, the accuracy of this result was debatable (Kemp et al., 2001). The overall test of conditional independence returned a value of 0.89, and since this was fairly close to 1.0, which would indicate conditional independence, it was felt that it was not necessary to reject the null hypothesis of conditional independence between the data themes (Hansen, 2000).

Once the premise of conditional independence had been accepted, it was then possible to evaluate the likelihood of a unit area containing a hominid site given the presence or absence of the evidential themes. In order to complete this calculation, the prior probability of a unit area containing a hominid site had to be known. In this instance, the prior probability of a hominid site occurring

Table 4. Chi - Squared (χ^2) Pairwise Test for Conditional Independence

	Carnivore/ Scavenger	Herbivore	Herbivore/Root/ Omnivore	Vegetation	Topography
Carnivore	0.67	0.02	5.10 *	2.27	0.02
Carnivore/ Scavenger		0.47	0.01	1.56	1.81
Herbivore			0.01	0.01	0.07
Herbivore/Root/ Omnivore				0.08	0.01
Vegetation					0.44

For every pairwise test df = 1, p=0.05, χ^2 =3.84
* Significant at 95%.

Table 5. Probability Table to Assess Conditional Independence

	Carnivore/ Scavenger	Herbivore	Herbivore/Root/ Omnivore	Vegetation	Topography
Carnivore	0.41	0.88	0.00	0.13	0.88
Carnivore/ Scavenger		0.49	0.94	0.21	0.18
Herbivore			0.91	0.93	0.80
Herbivore/Root/ Omnivore				0.77	0.91
Vegetation					0.51

Values < .05 indicate some conditional dependence
Overall Test of Conditional Independence = 0.89

Illustration 12. Posterior Probability Surface to Show Area of Possible Hominid Migration.

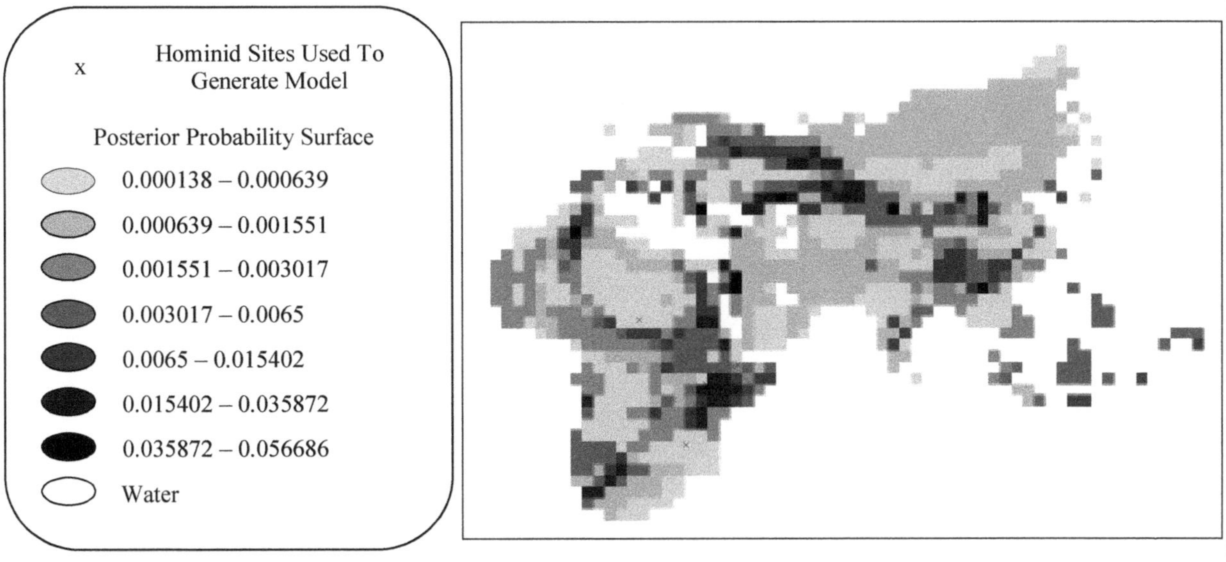

Illustration 13. Confidence Map To Show Areas Of Possible Hominid Migration. (Probability Surface of GIS Model Normalised by Total Uncertainty).

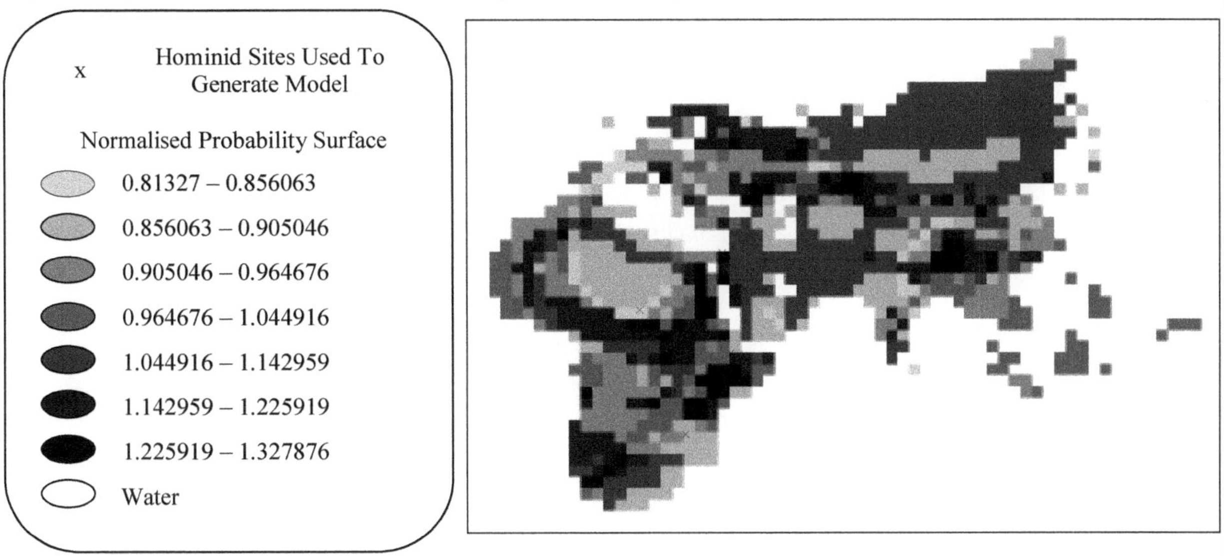

within a unit area was 0.0031, based on the number of hominid sites being 17 and the number of units within the research area being 5455.6. 'The posterior probability (or, as expressed in Arc-SDM, posterior logits where logits are natural logarithms) of each of the six data themes could then be calculated from the prior probability figure' (Kemp et al., 1999). One value or weight was returned if the theme was present in the unit area, and another if the theme was absent. The posterior logits for each of the themes in every unit area were then added and converted back into posterior probabilities to produce the response map expressing the probability that a unit area would contain a hominid site (Illustration 12) (Raines et al., 2000).

The results show that, based on the six independent variables used to generate the model, unit areas with higher posterior probability values than the prior probability of 0.0031 depict areas where the likelihood of hominid sites occurring is greater than would be expected to occur by chance. Conversely, unit areas having a smaller posterior probability value than prior probability value have a lower probability of containing hominid sites than would be expected to occur by chance. It was also possible to produce a map illustrating the amount of confidence in the posterior probability map by normalising the posterior probability values by the amount of total uncertainty in the data. This confidence map is shown in Illustration 13.

10.3.4 Testing the Validity of the Model

One of the most important aspects of generating the GIS model was to determine the validity of the model results; this procedure is discussed more fully in Section 3. One procedure to test whether the model was capable of predicting the location of hominid sites was to see whether non-site localities were found predominantly in areas of low probability for hominid sites. Owing to the sparsity of data, it was not possible to know for certain the position of non-site locations, but the fact that the model highlighted areas of desert and rainforest as being unlikely areas for hominid migration conformed with the current state of knowledge.

Another way of assessing the accuracy of the predictions was to use data independent from the training sample used to produce the model, and to this end hominid sites from South Africa were omitted from the initial model generation. In order to make this assessment more comparable to the rest of the research area, known faunal evidence from this area was also excluded. This was due to the fact that over much of the research area information relating to Plio-Pleistocene fauna was sparse and so the model relied upon interpolation from known faunal sites. Therefore, the probability of the occurrence of the South African hominid sites should also be based upon interpolated data. From the posterior probability response map, shown in Illustration 12, it can clearly be seen that the whole of the country falls within unit areas where the predicted probability of locating a hominid site is low, in fact lower than the prior probability of 0.0031; however, the level of confidence in this result is low to medium, see Illustration 13.

A final way of testing the validity of the model results was to use the training sample, which was the hominid site point data on which the model was constructed. The results of this analysis can be seen in Table 6. Nearly half the known hominid sites fell within the posterior probability intervals designated with a high level of probability for predicting hominid sites. Furthermore, 64.71% of sites fell within unit areas having a higher posterior probability value than prior probability value.

10.4 Discussion

The results clearly show that there was a pattern linking the location of early hominid sites with the distribution of flora, fauna and topography and that this model was capable of identifying certain areas where the probability of discovering additional hominid sites was better than that of chance.

Table 6. Validity of the Arc-SDM Model Based on the Training Sample Data.

Posterior Probability Interval	Code Number of Hominid Sites *	Number of Sites	Percentage of Total Sites	Level of Probability
0.000138 - 0.000639	6, 104	2	11.76	Low
0.000639 - 0.001551		0	0	Low
0.001551 - 0.003100	92, 35a, 4, 10	4	23.53	Low
0.003100 - 0.006500		0	0	Medium
0.006500 - 0.015402	98, 15, 51	3	17.65	Medium
0.015402 - 0.035872	21, 103, 80, 55, 81, 26, 62, 71	8	47.06	High
0.035872 - 0.056686		0	0	High

* Site names for each of these codes can be found in Table 7.

The Weights-of-Evidence method is a fairly new technique to have been developed to predict areas of potential location of the variable under consideration; be it rock types, consumer spending or archaeological sites. As outlined above, this procedure was the best way of generating a GIS model of early hominid migration since, unusually for most models, it did not require the addition of non-site data. Whilst logistic regression and discriminant analysis are probably the most commonly used methods of determining site location within archaeological scenarios, their use in this instance would have been statistically incorrect since a database of non-site environmental data could not have been prepared (Savage, 1990).

The small number of sites from which relevant faunal remains have been recovered affected the accuracy of the model, as can be seen in the failure of the model to identify South Africa as a suitable area for the location of hominid sites. However, the benefit of the Arc-SDM method was that the complementary confidence map also produced showed that the results obtained for this area could not be taken at face-value. The amount of variance encountered in the production of the weights implied that the results obtained could only be accepted at a fairly low confidence level. By comparison, the East Africa Rift Valley system returned a high probability for the location of hominid sites with a correspondingly high level of confidence in the results. Whilst this level of confidence is likely to be slightly inflated, since the sites in this area also represented the training sample, it does provide an indication of what could be achieved if the level of knowledge regarding the distribution of large mammalian faunal species over the rest of the research area was increased.

When the results of the probability surface map and the confidence map (Illustrations 12 & 13) were analysed in conjunction, a reliable indication of hominid migration patterns was identified. The probability map highlighted the East African Rift Valley system as one of the areas most likely to contain early hominid sites with a corridor extending in a southwesterly direction through Zambia and Zaire to another area with a reasonable expectation of containing hominid sites in Botswana and Namibia. The level of confidence in these results was quite high to high. One of the existing hominid sites known in this area, that of Uraha in the Malawi Rift Valley appeared to confound this pattern since it was found in an area with a predicted low probability of hominid sites. As with the South African sites, however, the level of confidence returned for this result was extremely low, further confirming the necessity of interpreting the probability surface in association with the confidence map.

Moving northwards, the Sudan represented another area where the probability of hominid site occurrence was better than would be expected, and this pattern continued around the east, south and west sides of the Sahara. The central Saharan region and the rainforest and deciduous forest regions of Plio-Pleistocene Africa were all identified as areas with extremely low probabilities of hominid occurrence and the levels of confidence were sufficient to confirm these results in all areas except the Sahara. Here, the level of confidence in the results was generally low and extremely low in the region of the hominid site of Bahr el Ghazal, Chad (Brunet et al., 1995; Zazzo et al., 2000). This reflected a large variability in the weights returned for the area, primarily the contradiction of the location of the site within a desertic environment but with the recorded presence of hominids and other fauna. As has been discussed in some detail above, (Section 8.1), the Sahara varied extensively in size and extent during the Plio-Pleistocene and it is likely that at certain times the margins would have been suitable for hominid migration whilst at other times they would not.

The probability surface and confidence level results implied that there was no passageway for hominid migration across the Straits of Bab el Mandeb and into Arabia during the Late Pliocene and Early Pleistocene. However, this result may have been affected by the identification of desert in this area in the Pliocene vegetation dataset, a fact that is open to doubt (see Section 8.3). Additional clarification of the data recorded from this area is therefore necessary before the rejection of the southern Red Sea hominid migration route can be made with any certainty.

The Levantine route for hominid migration out of Africa appeared to represent the most likely scenario from the results obtained from this model. Both the probability surface and confidence level results were in accord; however, it was not clear in what direction further migrations were most likely. The vast majority of central Asia including Syria, Iran, Iraq, Afghanistan, Pakistan and southern Russian republics all returned probability values of locating evidence for hominid migration as less than that occurring by chance and, on the whole, the confidence results confirmed this with some certainty. Nevertheless, the model predicted that hominid migration towards eastern Asia had to have occurred since areas of southern China were envisaged, with a high level of confidence, as being fairly likely to contain early hominid sites.

This migration towards the east may have taken place via a long and circuitous route. The model showed that there was a slight possibility that hominid migrations out of Africa may have occurred via the Straits of Gibraltar, with migration subsequently following an easterly direction along the northern coast of the Mediterranean. This routing is difficult to interpret however, as the rasterisation of the data has caused the coastline to appear extremely 'blocky', with some land categorised as water and vice versa. It may be necessary to examine the data from these regions in more detail before a definite prognosis can be made. Even so, the discovery of Lower Pleistocene hominid remains from the Iberian Peninsula (e.g. Carbonell et al., 1995; Gibert et al., 1998; Martínez Navarro et al., 1997; Palmqvist, 1997; Parés and Pérez-González, 1995; Roe, 1995) represent reasons to remain optimistic.

The model predicted, with some confidence, that further easterly hominid migration was likely to have occurred through the more northerly countries of central Asia, specifically Georgia, Ukraine, Russia and Kazakhstan. Currently there is no evidence to suggest that the limits of the Black and Caspian Seas were substantially different in Plio-Pleistocene than in the modern world. Recent research into the palaeogeography and palaeoclimatology of the area (e.g. Borisova, 1991; Borisova, 1994; Grichuk, 1991; Iossifova and Semenov, 1998; Svetlitskaya, 1994) gives no reason to suppose that much of the region would have been uninhabitable during the period in question, in fact it could have acted as a refugium. Research in these countries has been somewhat limited over recent years, however, the discovery of the hominid remains at Dmanisi, Georgia confirms that further work could prove profitable.

Unfortunately, many of the countries listed above extend, either in part or entirely, above 40° of latitude where the projection of the data is less accurate than at slightly more southern latitudes. This would suggest that this corridor of predicted hominid migration was actually located slightly to the south, but this would bring in the added complication of negotiating the Caspian Sea. If this passageway was located far enough south to encompass the northern limits of Iran, this would appear to be the most logical route for hominids to have migrated along. Examination of the Pliocene vegetation map, Illustration 3, showed that this path through grassland vegetation belts would have avoided the patches of desert present in central Asia at the time and would have led to the area in southern China where the model predicted the location of hominid sites. Also, current levels of knowledge relating to the Plio-Pleistocene environment in Iran are extremely poor, and so this region should not necessarily be excluded from migration pathways. It would seem likely, therefore, that a migration route existed across central Asia to China during the Plio-Pleistocene period, but the origin of this route requires further confirmation.

One Chinese site in particular, that of Longgupo Cave (Huang Wanpo et al., 1995; Huang Wanpo et al., 1991) has been claimed to show evidence for hominid migration far to the east during the Plio-Pleistocene. On the other hand, much of the dating evidence from the site is not perceived as 100% credible by most Western academics, mostly due to the standard of excavation. Concern has also been raised over the identification of the hominid remains and the flaked stone tools also recovered. Nevertheless, existing flora and faunal data from China would suggest that, as the model predicts, at some point in the future more concrete evidence for hominid migration into the region will be recovered.

For several reasons it was not possible to examine the probability of hominid migration towards the Indonesian islands with this model. Firstly, the rasterisation of the data meant that the distribution of land and sea in the area was no longer a true depiction of the region. Secondly, the interpolated values for the faunal data themes were not accurate towards the extremes of the study area and lastly the necessary use of a projection meant that precision was compromised at the edges of the map. The last two factors also explain why it was not possible to interpret the results obtained from northern Europe, Russia and Siberia.

10.5 Conclusions

This inductive Arc-SDM model generated to simulate early hominid colonisation across Eurasia returned some significant results and represents a successful attempt to comprehend some of the restrictions that would have affected hominid migration. Whilst the model has some limitations, most notably relating to a lack of suitable data, subsequent discoveries of floral, faunal and hominid data can be added to the database and used to perfect the model. As such, the simulation is an invaluable addition to the corpus of information relating to hominid evolution.

One of the great benefits of these kind of predictive models is that potentially successful areas for new fieldwork are highlighted, such as in the Asian grasslands and southern China, and the type of information that should be collected in order to enhance further the model is already known. As was initially shown in Illustration 1, the uneven distribution of Plio-Pleistocene faunal remains across the research area and the vast distances between the data points are likely to have affected the model results. There is no doubt that this model would benefit from substantially more faunal data, particularly in the regions of central and eastern Asia, so as to increase the reliability of the interpolated data themes. The collection of this type of data would represent the next step in the refinement of the model.

Many of the migration pathways suggested by this model have already been hypothesised by various researchers; c.f. the importance of Asian grasslands to hominid migration (Dennell, 2001); crossing of the Straits of Gibraltar or Bab el Mandeb by macromammals (Arribas & Palmqvist, 1999; Delany, 1989; Martínez Navarro & Palmqvist, 1995, 1996; Straus, 2001). This shows that the model can be used to complement more traditional methods of archaeological research and can perhaps also be used to confirm or refute other theories of hominid migration.

Chapter 11

Statistical Analysis

As well as generating the GIS predictive model, it was decided to conduct statistical analysis of the faunal data collected in order to ascertain whether the two different methods for determining possible hominid location outside Africa achieved the same results. However, by its very nature this analysis was more restricted, since it was only possible to suggest that a site may have been inhabited by hominids once a faunal assemblage had been collected. Therefore, it was only possible to propose that more work be conducted in certain areas or sites rather than suggest new regions for exploratory work. As was discussed in the previous section and Section 5, further analysis of floral data was not attempted due to the inherent inaccuracies and problems associated with the data type.

11.1 Materials and Methods

11.1.1 Materials

The faunal data used in the statistical analyses was obtained from the literature, with Table 2 listing the sources of the data; this data was the same as that used to generate the GIS predictive model. The different macromammalian genera that had been recovered from each site were recorded and sites for which no faunal data were obtained were omitted from further analysis, since inclusion of these sites would skew any results (Fieller, pers. comm. 2001). It was found that many different genera were present at the 99 sites from which faunal remains had been recovered, with little overlap between the sites. That is, most genera were only found at one or two sites, meaning that, as with the GIS analysis, any analysis undertaken to determine whether particular genera were preferentially associated with hominids would produce meaningless results. Consequently, the data were once again incorporated into the larger categories of herbivore, herbivore/root, carnivore, carnivore/scavenger, omnivore and omnivore/scavenger diet type.

11.1.2 Methods

In order to produce a statistically robust analysis that would help determine at what sites it was likely that early hominids could have survived, hominid genera were removed from further study. Since only hominids had been classified in the omnivore/scavenger diet type, this category was omitted leaving five categories of diet type to be investigated. Table 7 shows the sites from which data used in this analysis was obtained as well as the number of genera falling within each diet type grouping.

A brief examination of Table 7 was sufficient to show that there was a marked difference both within and between sites in terms of the total number of genera represented and their distribution across the diet type categories, with herbivores apparently substantially more abundant than carnivores and omnivores. This observation is confirmed in Table 8, where the mean and standard deviation values for each class are listed. Nevertheless, it was decided that the data should not be standardised before conducting statistical analysis since higher numbers of herbivores would be expected as they tend to reside in large herds whilst carnivores are more solitary animals. Also, the total size of the assemblage was a real feature of the data and not simply a reflection of the amount of work conducted at the site and so the use of the actual numbers would produce more accurate results (Fieller, pers. comm. 2001). No other relationships were discernible.

In order to assess what statistical procedure would be most appropriate for the data, a correlation matrix for the diet type variables was produced, as illustrated in Table 9. This showed that there was a certain amount of correlation between the different variables, although omnivore was not well-correlated with any other diet type. This suggested that Principal Components Analysis (PCA) may be a suitable statistical technique for the data (SPSS for Windows, 2000), because the more closely related the original variables, the better the new summary variables would represent the data (Baxter, 1994; Shennan, 1997). This hypothesis was supported by the low significance levels for the correlations, also illustrated in Table 9 (SPSS for Windows, 2000).

Due to the extremely low abundance of omnivore genera, the restriction of this diet type to *Ursus* (see Appendix 1) and the apparent lack of omnivores within Africa, it was felt that a slightly different classificatory scheme would reflect the data more accurately. Genera that had been included within the herbivore/root diet type category were restricted to the families Anthracotheriidae and Suidae, and an argument can be made for their inclusion within the omnivore diet type. This led to a revision of the diet type groupings to herbivore, carnivore, carnivore/scavenger and combined herbivore/root and omnivore. Table 10 shows the descriptive statistics and Table 11 the correlation matrix for these revised categories. The revised groupings led to little change from the original data in both the correlation and significance results. There remained a certain amount of correlation between the different variables, denoting once again that PCA would be a suitable statistical technique. The low significance levels for the correlations also corroborated this.

It was decided, therefore, to undertake Principal Component Analysis on both the original and revised diet type groupings utilising the statistical package SPSS version 10.0.7.

Table 7. Count of Genera Falling Within Each Diet Type Category for Every Site Used in the Principal Components Analysis.

Site Code	Site	Count of Genera of Each Diet Type: Herbivore	Herbivore/ Root	Omnivore	Combined Herbivore/ Root & Omnivore	Carnivore	Carnivore/ Root
1	Africa	43	10	0	10	4	4
2	Ahl Al Oughlam Quarry	21	2	1	3	4	2
3	Apollonia, C. Macedonia	5	0	0	0	0	0
4	Aramis	0	0	0	0	1	0
5	Atapuerca - TD6	6	1	1	2	1	1
6	Bahr el Ghazal	8	1	0	1	0	0
7	Betfia 2, Crisana	0	0	1	1	7	1
8	Bethlehem	6	1	0	1	2	0
9	Bijiashan, Liuzhou, Guangxi	0	0	0	0	1	0
10	Bouri Peninsular, S. Afar	25	3	0	3	1	0
12	Central Europe	0	0	1	1	7	2
13	Chambal Ridge	5	0	0	0	0	0
14	Chandigarh, Pinjore, Upper Siwalik	16	2	0	2	1	0
15	Chemeron Formation	19	3	0	3	0	1
16	Chiaijiashan, Tangshan, Heibi	3	0	0	0	1	0
18	Daferno, W. Macedonia	8	0	0	0	0	0
19	Damiao, Wushan, Sichuan	7	1	0	1	1	1
20	Deutsch - Altenburg 2, Basse - Autriche	0	0	1	1	6	0
21	Dmanisi	14	1	1	2	4	1
22	Dongpaoshan, N.W. Hunan	1	0	0	0	0	0
25	Europe	44	2	1	3	8	3
26	Fejej	6	3	0	3	0	0
27	Fuenta Nueva - 3, Orce	9	0	1	1	1	0
29	Gaopen, Jianshi, Hubei	5	0	0	0	0	1
30	Gaoping, W. Hubei	1	0	0	0	0	0
31	Gerakarou, C. Macedonia	6	1	0	1	0	0
32	Gesher Benot Ya'aqov	7	0	0	0	0	0
33	Gonghe (lower)	1	0	0	0	0	0
34	Grotte Zabia, Petite - Pologne	0	0	1	1	1	0
35	Hadar. All Flora. Fauna 4.0 - 2.5 Ma	17	4	0	4	2	1
35a	Hadar. Fauna 2.5 - 2.0 Ma	15	3	0	3	2	1
35b	Hadar. Fauna Middle Pleistocene.	12	3	0	3	0	0
37	Himachal Pradesh	13	0	0	0	0	0
39	Holstein, Moravie	0	0	1	1	1	0
40	Houhecun, Dali, Shaanxi	0	0	0	0	1	0
45	Jing - Le Basin, Shanxi Province	6	0	0	0	0	0
48	Jockgrim, Palatinat	0	0	1	1	1	0
49	Juyuangong	1	0	0	0	0	0
50	Kamyk, Petite - Pologne	0	0	0	0	3	0
51	Kanjera Formation	30	3	0	3	1	1
54	Kokiselei	5	0	0	0	0	0
55	Koobi Fora	41	4	0	4	7	3
56	Krimni - 1, C. Macedonia	3	0	0	0	0	0
58	Lantian, Shaanxi	11	1	1	2	3	1
59	Latamne	12	0	0	0	1	1
61	Lengzhaishan, Loucheng, Guangxi	7	1	0	1	0	1
62	Lokalalei	10	2	0	2	0	0
63	Longgupo (lower)	1	0	0	0	0	0
64	Lower Karewa, Pinjor of Upper Siwaliks	6	0	0	0	0	0
65	Mangla - Samwal Anticline	3	0	0	0	0	0
66	Meiningen, Thuringe	0	0	0	0	3	0
69	Mosbach 1, Hesse	0	0	1	1	0	1
70	Nagharsányhegy 2	0	0	1	1	2	0
71	Naiyena Engol	2	0	0	0	0	0
73	North Africa & Middle East	12	2	0	2	3	0
74	North Asia	10	0	1	1	4	1

Table 7 cont. Count of Genera Falling Within Each Diet Type Category for Every Site Used in the Principal Components Analysis.

Site Code	Site	Count of Genera of Each Diet Type Herbivore	Herbivore/ Root	Omnivore	Combined Herbivore/ Root & Omnivore	Carnivore	Carnivore/ Root
75	North China	2	0	0	0	1	0
76	North China Region	30	1	1	2	12	2
77	North of Qinling Mountains	8	0	0	0	1	1
79	Old World	1	0	0	0	3	2
80	Olduvai Gorge. 2.5 - 1.9 Ma	26	6	0	6	10	1
80a	Olduvai Gorge. 1.7 - 0.8 Ma	38	9	0	9	7	4
81	Omo. 3.0 - 1.9 Ma	38	6	0	6	8	4
81a	Omo. 1.9 - 1.0 Ma	35	5	0	5	6	2
82	Osztramos 2	0	0	1	1	3	0
83	Pabbi Hills	6	0	0	0	0	0
87	Ravin of Voulgarakis, C. Macedonia	1	0	0	0	0	0
88	Rohtas Anticline	2	0	0	0	0	0
92	Semliki Valley, Lusso Beds	14	3	0	3	1	0
94	South Asia	24	3	1	4	7	1
95	South China	4	0	0	0	1	1
96	South China Region	29	3	1	4	9	1
98	Sub - Saharan Africa	21	4	0	4	6	2
99	Tatrot	2	0	0	0	0	1
100	Tianqiao	1	0	0	0	0	0
101	Transitional Region, China	17	1	1	2	5	1
103	'Ubeidiya	15	2	1	3	6	1
104	Uraha, Karonga, Malema, Chiwondo Beds	37	5	0	5	1	0
105	Vassiloudi, C. Macedonia	2	1	0	1	0	0
106	Villány 5	0	0	1	1	3	0
107	W. Asia & Europe	19	1	1	2	8	1
109	Yinshan (lower)	1	0	0	0	0	0
110	Yuanmou Formation	16	1	0	1	4	1
111	Zoukoudian	1	0	0	0	0	1
112	Upper Siwaliks 1.8 - 1.6 Ma	0	0	0	0	2	1
112a	Upper Siwaliks 1.4 - 1.2 Ma	0	0	1	1	4	3
113	Site 642, Punjab. Guijrat District	5	1	0	1	0	1
116	Volax	1	0	0	0	4	0
117	Drama Basin	1	0	0	0	0	0
118	Serres	1	0	0	0	0	0
119	Neapolis	9	0	1	1	1	1
120	Tsotyli	1	0	0	0	0	0
121	Ptolemais	1	0	0	0	0	0
122	Ioannina	1	0	0	0	0	0
123	Elis	1	0	0	0	0	0
124	Pyrgos	3	0	0	0	1	0
125	Vlachiotis	1	0	0	0	0	0
126	An Nafud	8	0	0	0	2	1
127	Slivnitsa	12	0	0	0	4	1

Table 8. Descriptive Statistics for the Five Original Diet Type Categories.

Descriptive Statistics		
	Mean	Standard Deviation
Herbivore	9.27	11.33
Herbivore/Root	1.07	1.91
Carnivore	2.07	2.75
Carnivore/Scavenger	0.64	0.95
Omnivore	0.25	0.44

Table 9. Matrix of Correlations and Levels of Significance Between the Five Original Variables Defining Diet Type.

Correlation Matrix						
		Herbivore	Herbivore/ Root	Carnivore	Carnivore/ Scavenger	Omnivore
Correlation	Herbivore	1.00	0.83	0.60	0.67	0.05
	Herbivore/Root	0.83	1.00	0.45	0.62	-0.11
	Carnivore	0.60	0.45	1.00	0.64	0.48
	Carnivore/Scavenger	0.67	0.62	0.64	1.00	0.20
	Omnivore	0.05	-0.11	0.48	0.20	1.00
Significance. (1-tailed)	Herbivore		0.00	0.00	0.00	0.30
	Herbivore/Root	0.00		0.00	0.00	0.14
	Carnivore	0.00	0.00		0.00	0.00
	Carnivore/Scavenger	0.00	0.00	0.00		0.02
	Omnivore	0.30	0.14	0.00	0.00	

Table 10. Descriptive Statistics for the Revised Diet Type Categories.

Descriptive Statistics		
	Mean	Standard Deviation
Herbivore	9.27	11.33
Carnivore	2.07	2.75
Carnivore/Scavenger	0.64	0.95
Omnivore & Herbivore/Root	1.32	1.91

Table 11. Matrix of Correlations and Levels of Significance Between the Revised Variables Defining Diet Type.

Correlation Matrix					
		Herbivore	Carnivore	Carnivore/ Scavenger	Omnivore & Herbivore/Root
Correlation	Herbivore	1.00	0.60	0.67	0.84
	Carnivore	0.60	1.00	0.64	0.56
	Carnivore/Scavenger	0.67	0.64	1.00	0.67
	Omnivore & Herbivore/Root	0.84	0.56	0.67	1.00
Significance. (1-tailed)	Herbivore		0.00	0.00	0.00
	Carnivore	0.00		0.00	0.00
	Carnivore/Scavenger	0.00	0.00		0.00
	Omnivore & Herbivore/Root	0.00	0.00	0.00	

11.2 Results

PCA is an important tool for obtaining a picture of complex multivariate data since it allows a structure of the data to be seen that is not obvious from the original data matrix, providing a mechanism to explain how different variables are related (Baxter, 1994). As has been already observed, an examination of Table 7 failed to elucidate any obvious patterning in the data, implying that multivariate statistical techniques would be required. Whilst it would have been possible to construct several bivariate scatterplots of the data arranged as a matrix plot, it is likely that the results would have been difficult to interpret accurately (Fieller, 2001). PCA, as a data reduction technique, allows any interpretation to be made more easily since its main objective is to portray as much variation as possible shown by the data in two dimensions (Baxter, 1994). This is achieved by producing new variable axes from the matrix of correlation coefficients that are orthogonal (at 90° and therefore unrelated to each other), and still contain much, if not all, of the information held in the original data (Fieller, 2001; Shennan, 1997). A scatterplot constructed on the new axes shows the data in a more understandable manner allowing conclusions to be drawn (Fieller, 2001).

Table 12 shows the results of the PCA run on the five original diet type categories. The first component accounted for 59.27% of the variance in the original data, the second for 25.01% and the third for 7.14%; with a cumulative total of 91.42% of the variance in the original data accounted for by the first three principal components. The remaining two principal components are responsible for explaining little of the original variation and can therefore be safely omitted from further discussion (Fieller, pers. comm. 2001).

Table 12. Eigenvalues and Percentages of Variance Accounted for by the Five Principal Components of the Original Diet Type Data. Only the First Three Components Accounted for Substantial Proportions of the Variance in the Original Data.

Total Variance Explained

Component	Initial Eigenvalues			Extraction Sums of Squared Loadings		
	Total	% of Variance	Cumulative %	Total	% of Variance	Cumulative %
1	2.96	59.27	59.27	2.96	59.27	59.27
2	1.25	25.01	84.28	1.25	25.01	84.28
3	0.36	7.14	91.42	0.36	7.14	91.42
4	0.28	5.65	97.06			
5	0.15	2.94	100.00			

The component matrix, illustrated in Table 13, allows a more in-depth interpretation of the results of the analysis (Shennan, 1997). The first component is a reflection of the size of the assemblages being examined, that is the overall abundance, seen in the fact that the loadings of the variables on this component all have the same sign and, apart from omnivores, are all of a similar size (Baxter, 1994; Fieller, 2001). This means that sites with a large first principal component have many different genera within the assemblage, whilst those with a small first principal component have few different genera within the assemblage (Fieller, pers. comm. 2001). The second principal component represents a clear distinction between the herbivorous and other diet type categories, with omnivores standing out as being of particular importance in defining this variable. Therefore, the second principal component is a measure of the dominance of herbivores or omnivores within each assemblage, with high values representing large numbers of omnivores and few herbivores and vice versa. Finally, the third principal component reflects the ratio of carnivores and carnivore/scavengers to omnivores and the herbivorous diet type groupings, with carnivore/scavengers being of particular importance in discriminating the data. High values of this component indicate larger numbers of omnivore, herbivore and herbivore/root genera with the carnivorous diet types being less abundant, and vice versa.

Table 13. Loadings of the Original Diet Type Variables on the First Three Principal Components.

Component Matrix

	Component		
	1	2	3
Herbivore	0.90	-0.24	0.22
Herbivore/Root	0.83	-0.44	0.22
Carnivore	0.81	0.40	-0.03
Carnivore/Scavenger	0.86	0.02	-0.48
Omnivore	0.26	0.91	0.18

These conclusions can be visualised more easily when the results are portrayed in graphical form, as in Illustrations 14, 15 and 16. Illustration 14 is a plot of the loadings of the original variables on the first two principal components, as shown in Table 13. This demonstrates very clearly that the omnivore diet type is distinctly

Illustration 14. Scattergram of the Loadings of the Original Diet Type Variables On the First Two Principal Components.

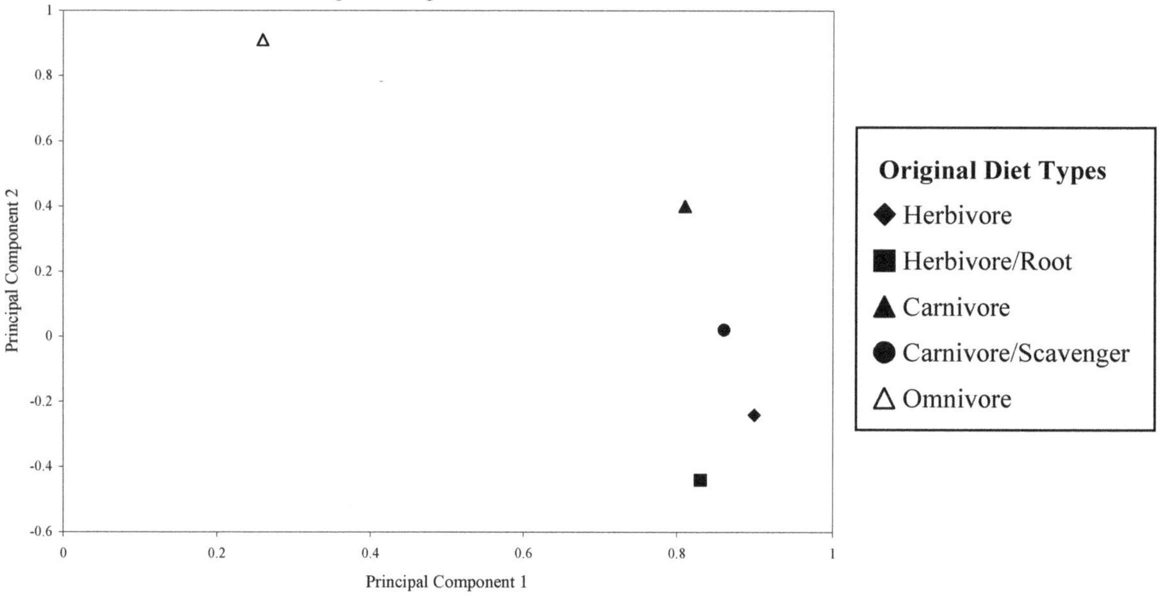

different from the other categories, primarily due to its low abundance in this data. The second principal component is of more use than the first in differentiating between the different diet types, although once again the omnivore category stands out as an anomaly, suggesting that the decision to alter the diet type categories and re-run the analysis was correct.

Illustration 15 is a plot of the individual site scores on the first two principal components from the original diet type groupings, as listed in Table 14. It is immediately apparent that the data is divided into two distinct sets, with further subsets also visible. As could be predicted from the previous information, the most obvious difference is as a result of the presence/absence of omnivore genera although, interestingly, there would also appear to be clustering of sites based on the presence/absence of hominids and herbivore/root genera.

Illustration 15. Scattergram of Individual Site Scores on the First Two Principal Components Based on the Original Diet Type Categories.

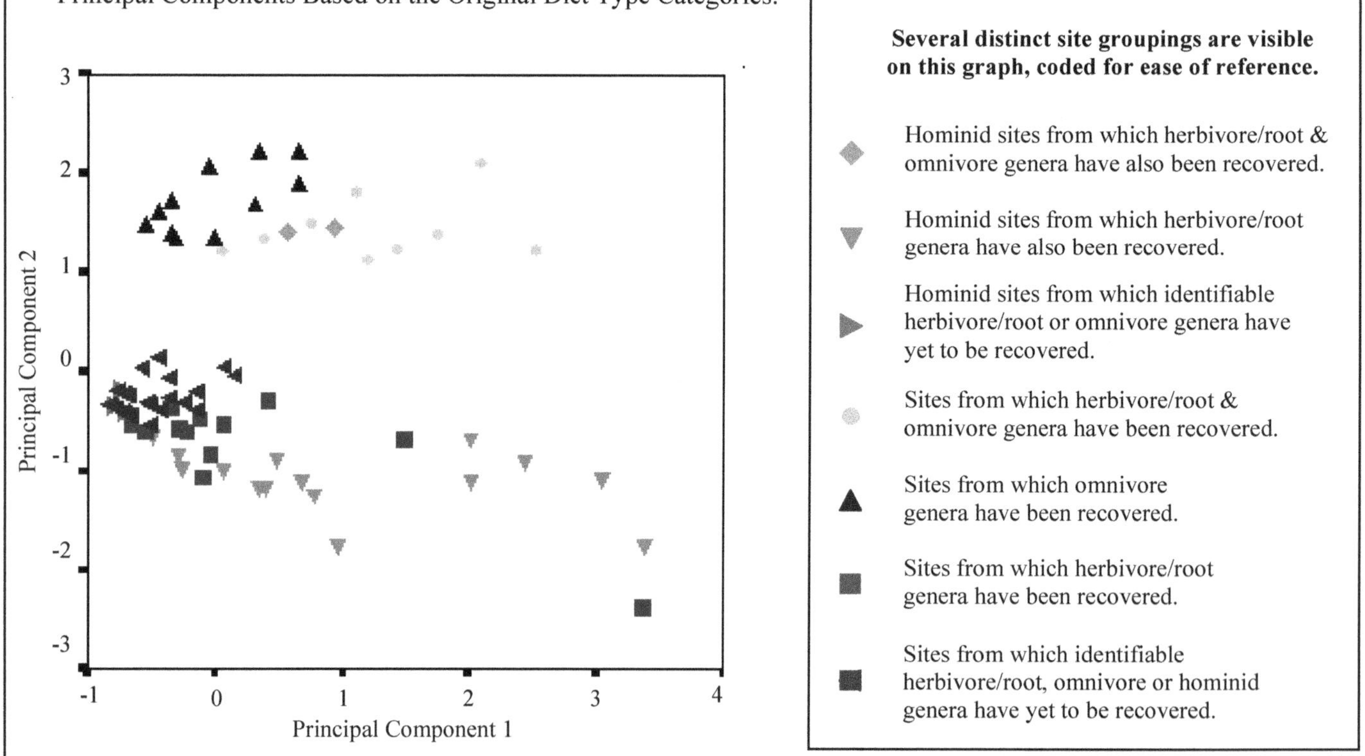

Illustration 16. Scattergram of Individual Site Scores on the Second and Third Principal Components Based on the Original Diet Type Categories.

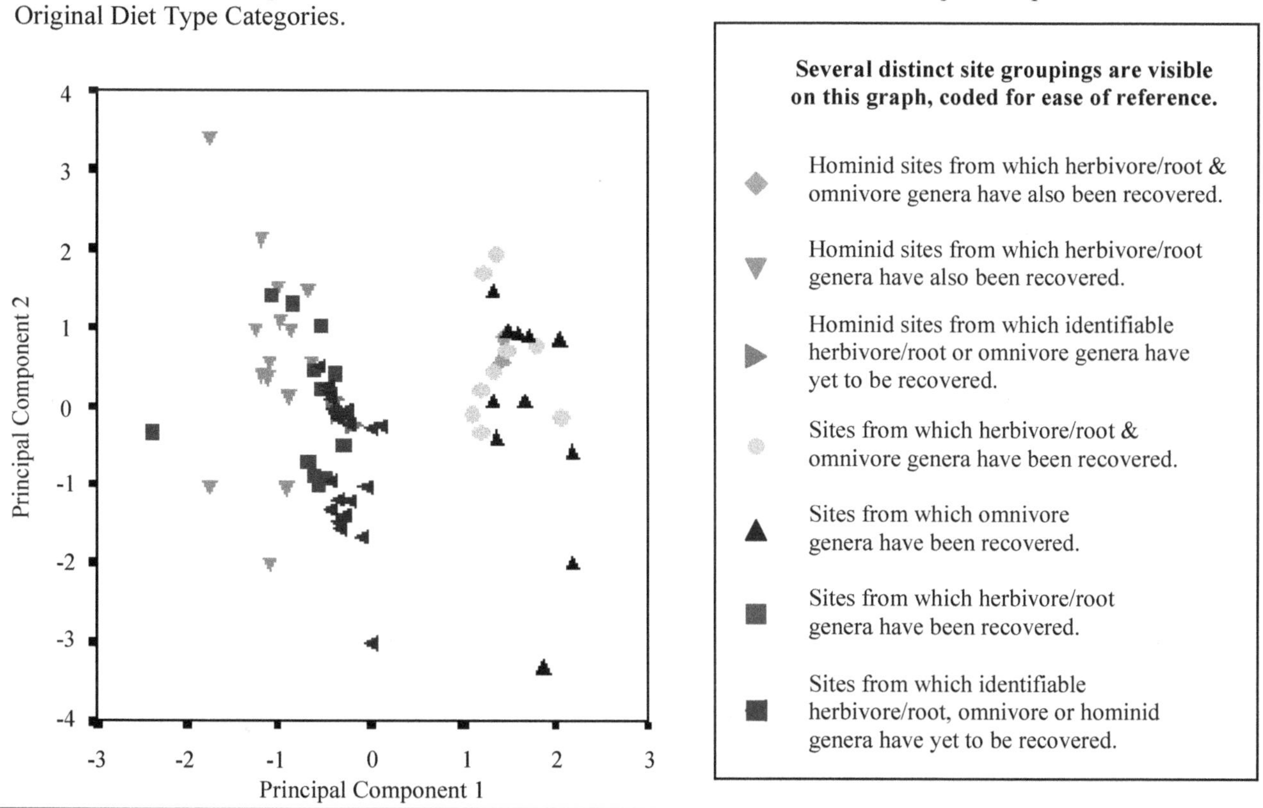

Illustration 16 is a plot of the individual site scores on the second and third principal components from the original diet type groupings, as listed in Table 14. As component 2 is once again plotted, the same bilateral division of the sites into those with omnivore genera and those with herbivore genera is repeated. Component 3 is responsible for some additional separation of the data on the basis of the presence or absence of carnivore genera. However, this additional component appears to have failed to elucidate any further patterning of the data than was visible on the plot of the first two components, perhaps not surprisingly since it only accounts for 7.14% of the original data variation.

Table 14. Principal Component Scores for the Original and Revised Diet Type Categories for Every Site Used in the Analysis.

Site Code	Site	Original Principal Component			Revised Principal Component		
		1	2	3	1	2	3
1	Africa	3.376	-2.373	-0.334	3.453	-2.864	-2.189
2	Ahl Al Oughlam Quarry	1.211	1.123	-0.133	1.172	0.021	-0.863
3	Apollonia, C. Macedonia	-0.723	-0.404	0.069	-0.713	-0.320	0.020
4	Aramis	-0.757	-0.200	-0.235	-0.749	0.304	0.148
5	Atapuerca - TD6	0.059	1.201	0.190	0.024	-0.361	-0.890
6	Bahr el Ghazal	-0.496	-0.639	0.553	-0.477	-0.821	0.259
7	Betfia 2, Crisana	0.349	2.191	-0.614	0.289	2.558	0.561
8	Bethlehem	-0.351	-0.370	0.391	-0.336	0.034	0.788
9	Bijiashan, Liuzhou, Guangxi	-0.757	-0.200	-0.235	-0.749	0.304	0.148
10	Bouri Peninsular, S. Afar	0.352	-1.182	2.108	0.386	-1.986	1.410
12	Central Europe	0.654	2.203	-2.011	0.590	2.895	-0.956
13	Chambal Ridge	-0.723	-0.404	0.069	-0.713	-0.320	0.020
14	Chandigarh, Pinjore, Upper Siwalik	-0.036	-0.843	1.290	-0.010	-1.190	0.967
15	Chemeron Formation	0.396	-1.182	0.402	0.430	-1.732	-0.609
16	Chiaijiashan, Tangshan, Heibi	-0.677	-0.252	-0.068	-0.669	0.157	0.251
18	Daferno, W. Macedonia	-0.642	-0.456	0.235	-0.633	-0.467	0.122
19	Damiao, Wushan, Sichuan	-0.118	-0.492	-0.925	-0.105	-0.056	-0.993
20	Deutsch - Altenburg 2, Basse - Autriche	-0.056	2.061	0.809	-0.108	1.842	1.778
21	Dmanisi	0.572	1.414	0.558	0.528	0.381	0.278
22	Dongpaoshan, N.W. Hunan	-0.830	-0.335	-0.154	-0.819	-0.123	-0.116

Table 14 cont. Principal Component Scores for the Original and Revised Diet Type Categories for Every Site Used in the Analysis.

Site Code	Site	Original Principal Component 1	2	3	Revised Principal Component 1	2	3
25	Europe	2.530	1.206	-0.355	2.474	0.742	-0.401
26	Fejej	-0.257	-0.971	1.078	-0.218	-1.431	0.464
27	Fuenta Nueva - 3, Orce	-0.312	1.320	1.436	-0.353	-0.492	0.592
29	Gaopen, Jianshi, Hubei	-0.418	-0.391	-1.329	-0.412	0.018	-1.496
30	Gaoping, W. Hubei	-0.830	-0.335	-0.154	-0.819	-0.123	-0.116
31	Gerakarou, C. Macedonia	-0.550	-0.604	0.442	-0.530	-0.723	0.191
32	Gesher Benot Ya'aqov	-0.669	-0.438	0.180	-0.659	-0.418	0.088
33	Gonghe (lower)	-0.830	-0.335	-0.154	-0.819	-0.123	-0.116
34	Grotte Zabia, Petite - Pologne	-0.553	1.475	0.936	-0.593	-0.050	0.285
35	Hadar. All Flora. Fauna 4.0 - 2.5 Ma	0.688	-1.097	0.558	0.726	-1.231	0.057
35a	Hadar. Fauna 2.5 - 2.0 Ma	0.488	-0.879	0.129	0.517	-0.779	-0.148
35b	Hadar. Fauna Middle Pleistocene.	-0.096	-1.074	1.411	-0.058	-1.726	0.669
37	Himachal Pradesh	-0.508	-0.542	0.513	-0.499	-0.713	0.293
39	Holstein, Moravie	-0.553	1.475	0.936	-0.593	-0.050	0.285
40	Houhecun, Dali, Shaanxi	-0.757	-0.200	-0.235	-0.749	0.304	0.148
45	Jing - Le Basin, Shanxi Province	-0.696	-0.421	0.124	-0.686	-0.369	0.054
48	Jockgrim, Palatinat	-0.553	1.475	0.936	-0.593	-0.050	0.285
49	Juyuangong	-0.830	-0.335	-0.154	-0.819	-0.123	-0.116
50	Kamyk, Petite - Pologne	-0.558	0.034	-0.285	-0.555	1.061	0.746
51	Kanjera Formation	0.791	-1.255	0.988	0.820	-1.894	0.064
54	Kokiselei	-0.723	-0.404	0.069	-0.713	-0.320	0.020
55	Koobi Fora	2.439	-0.900	-1.031	2.453	0.157	-0.665
56	Krimni - 1, C. Macedonia	-0.776	-0.369	-0.042	-0.766	-0.222	-0.048
58	Lantian, Shaanxi	0.392	1.349	0.417	0.351	0.150	-0.123
59	Latamne	-0.131	-0.395	-0.966	-0.128	0.052	-0.959
61	Lengzhaishan, Loucheng, Guangxi	-0.218	-0.609	-0.900	-0.202	-0.435	-1.292
62	Lokalalei	-0.296	-0.857	0.982	-0.267	-1.274	0.464
63	Longgupo (lower)	-0.830	-0.335	-0.154	-0.819	-0.123	-0.116
64	Lower Karewa, Pinjor of Upper Siwaliks	-0.696	-0.421	0.124	-0.686	-0.369	0.054
65	Mangla - Samwal Anticline	-0.776	-0.369	-0.042	-0.766	-0.222	-0.048
66	Meiningen, Thuringe	-0.558	0.034	-0.285	-0.555	1.061	0.746
69	Mosbach 1, Hesse	-0.348	1.371	-0.436	-0.389	-0.091	-1.530
70	Nagharsányhegy 2	-0.454	1.592	0.911	-0.496	0.328	0.584
71	Naiyena Engol	-0.803	-0.352	-0.098	-0.793	-0.173	-0.082
73	North Africa & Middle East	0.056	-0.540	1.017	0.077	-0.237	1.428
74	North Asia	0.318	1.666	0.018	0.265	0.932	0.005
75	North China	-0.704	-0.235	-0.123	-0.696	0.206	0.217
76	North China Region	2.102	2.087	-0.155	2.031	2.960	1.696
77	North of Qinling Mountains	-0.238	-0.326	-1.188	-0.235	0.249	-1.096
79	Old World	0.079	0.042	-3.025	0.073	1.687	-2.253
80	Olduvai Gorge. 2.5 - 1.9 Ma	2.018	-0.682	1.490	2.054	0.645	3.026
80a	Olduvai Gorge. 1.7 - 0.8 Ma	3.395	-1.752	-1.006	3.454	-1.129	-1.600
81	Omo. 3.0 - 1.9 Ma	3.055	-1.085	-1.985	3.083	0.312	-1.712
81a	Omo. 1.9 - 1.0 Ma	2.020	-1.110	0.377	2.051	-0.619	0.485
82	Osztramos 2	-0.354	1.709	0.886	-0.399	0.707	0.882
83	Pabbi Hills	-0.696	-0.421	0.124	-0.686	-0.369	0.054
87	Ravin of Voulgarakis, C. Macedonia	-0.830	-0.335	-0.154	-0.819	-0.123	-0.116
88	Rohtas Anticline	-0.803	-0.352	-0.098	-0.793	-0.173	-0.082
92	Semliki Valley, Lusso Beds	0.057	-0.992	1.497	0.093	-1.446	1.036
94	South Asia	1.431	1.226	1.673	1.398	0.317	1.788
95	South China	-0.345	-0.256	-1.410	-0.342	0.445	-1.232
96	South China Region	1.764	1.374	1.900	1.725	0.828	2.556
98	Sub - Saharan Africa	1.498	-0.685	-0.719	1.522	0.423	-0.129
99	Tatrot	-0.498	-0.339	-1.495	-0.492	0.165	-1.599
100	Tianqiao	-0.830	-0.335	-0.154	-0.819	-0.123	-0.116
101	Transitional Region, China	0.752	1.479	0.699	0.705	0.612	0.679
103	'Ubeidiya	0.944	1.448	0.881	0.905	0.734	1.046
104	Uraha, Karonga, Malema, Chiwondo Beds	0.966	-1.756	3.410	1.018	-3.284	2.093
105	Vassiloudi, C. Macedonia	-0.657	-0.535	0.220	-0.637	-0.527	0.055
106	Villány 5	-0.354	1.709	0.886	-0.399	0.707	0.882

Table 14 cont. Principal Component Scores for the Original and Revised Diet Type Categories for Every Site Used in the Analysis.

Site Code	Site	Original Principal Component 1	2	3	Revised Principal Component 1	2	3
107	W. Asia & Europe	1.104	1.796	0.734	1.049	1.649	1.643
109	Yinshan (lower)	-0.830	-0.335	-0.154	-0.819	-0.123	-0.116
110	Yuanmou Formation	0.421	-0.296	-0.502	0.425	0.637	0.210
111	Zoukoudian	-0.525	-0.322	-1.551	-0.519	0.214	-1.633
112	Upper Siwaliks 1.8 - 1.6 Ma	-0.353	-0.070	-1.657	-0.351	1.020	-1.069
112a	Upper Siwaliks 1.4 - 1.2 Ma	0.660	1.865	-3.332	0.600	2.098	-3.368
113	Site 642, Punjab. Guijrat District	-0.271	-0.574	-1.011	-0.256	-0.336	-1.360
116	Volax	-0.432	0.134	-0.255	-0.432	1.390	1.078
117	Drama Basin	-0.830	-0.335	-0.154	-0.819	-0.123	-0.116
118	Serres	-0.830	-0.335	-0.154	-0.819	-0.123	-0.116
119	Neapolis	-0.007	1.332	0.039	-0.052	-0.154	-0.925
120	Tsotyli	-0.830	-0.335	-0.154	-0.819	-0.123	-0.116
121	Ptolemais	-0.830	-0.335	-0.154	-0.819	-0.123	-0.116
122	Ioannina	-0.830	-0.335	-0.154	-0.819	-0.123	-0.116
123	Elis	-0.830	-0.335	-0.154	-0.819	-0.123	-0.116
124	Pyrgos	-0.677	-0.252	-0.068	-0.669	0.157	0.251
125	Vlachiotis	-0.830	-0.335	-0.154	-0.819	-0.123	-0.116
126	An Nafud	-0.138	-0.209	-1.213	-0.138	0.627	-0.797
127	Slivnitsa	0.168	-0.043	-1.042	0.163	1.187	-0.063

Table 15. Eigenvalues and Percentages of Variance Accounted for by the Four Principal Components of the Revised Diet Type Data. Only the First Three Components Accounted for Substantial Proportions of the Variance in the Original Data.

Total Variance Explained

Component	Initial Eigenvalues			Extraction Sums of Squared Loadings		
	Total	% of Variance	Cumulative %	Total	% of Variance	Cumulative %
1	3.00	74.93	74.93	3.00	74.93	74.93
2	0.51	12.78	87.71	0.51	12.78	87.71
3	0.34	8.39	96.10	0.34	8.39	96.10
4	0.16	3.90	100.00			

Table 15 shows the results of the PCA run on the four revised diet type categories. The first component accounted for 74.93% of the variance in the original data, the second for 12.78% and the third for 8.39%; with a cumulative total of 96.10% of the variance in the original data accounted for by the first three principal components. The remaining principal component is responsible for explaining little of the original variation and can therefore be safely omitted from further discussion (Fieller, pers. comm. 2001).

Table 16. Loadings of the Revised Diet Type Variables on the First Three Principal Components.

Component Matrix

	Component 1	2	3
Herbivore	0.91	-0.28	0.13
Carnivore	0.80	0.53	0.28
Carnivore/Scavenger	0.86	0.16	-0.48
Omnivore & Herbivore/Root	0.89	-0.35	0.09

As previously, the component matrix, illustrated in Table 16, allows a more in-depth interpretation of the results of the analysis (Shennan, 1997). The first component is still a reflection of the overall abundance of the assemblages being examined as the variables all have the same sign and are all of a similar size (Baxter, 1994; Fieller, 2001). Therefore, the same conclusion can be drawn as above; sites with a large first principal component have many different genera within the assemblage, whilst those with a small first principal component have few different genera within the assemblage (Fieller, pers. comm. 2001). The second principal component

Illustration 17. Scattergram of the Loadings of the Revised Diet Type Variables On the First Two Principal Components.

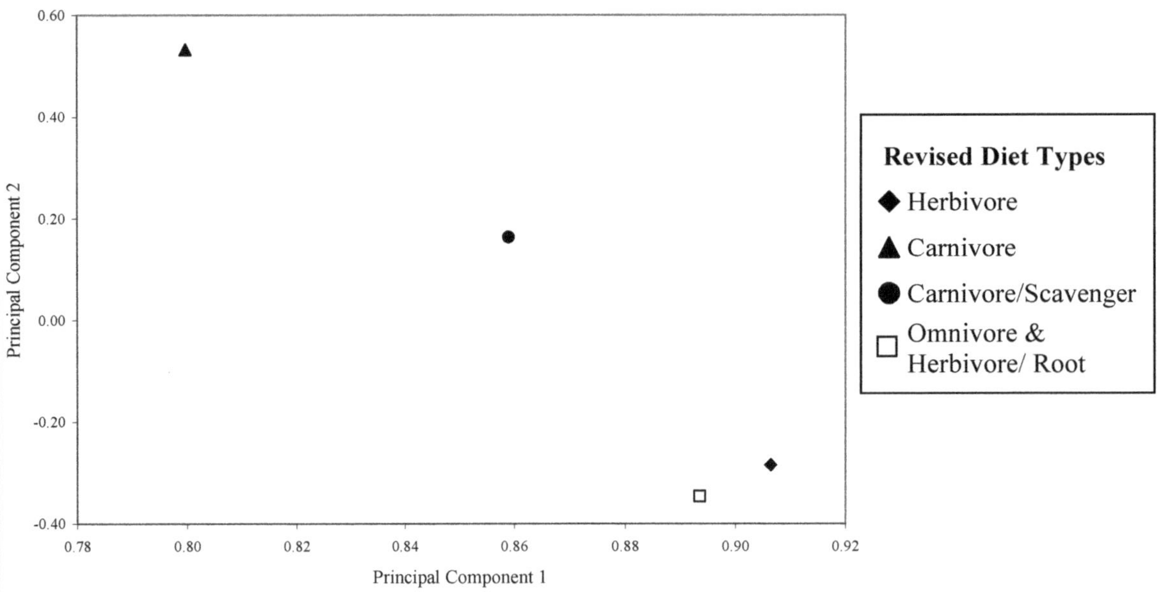

represents a clear distinction between the carnivorous and other diet type categories; with high values representing large numbers of carnivores and few other genera and vice versa. Finally, the third principal component reflects the importance of the carnivore/scavenger diet type. Low values of this component indicate larger numbers of carnivore/scavenger genera with the other diet types being less abundant, and vice versa.

Illustrations 17, 18 and 19 represent the results in graphical form. Illustration 17 is a plot of the loadings of the revised variables on the first two principal components, as shown in Table 16. It should be noted that whilst there is an apparent large amount of separation between the different diet types on component 1, the scale can be quite deceiving and in reality the results are little different from those obtained from the analysis on the original diet type

Illustration 18. Scattergram of Individual Site Scores on the First Two Principal Components Based on the Revised Diet Type Categories.

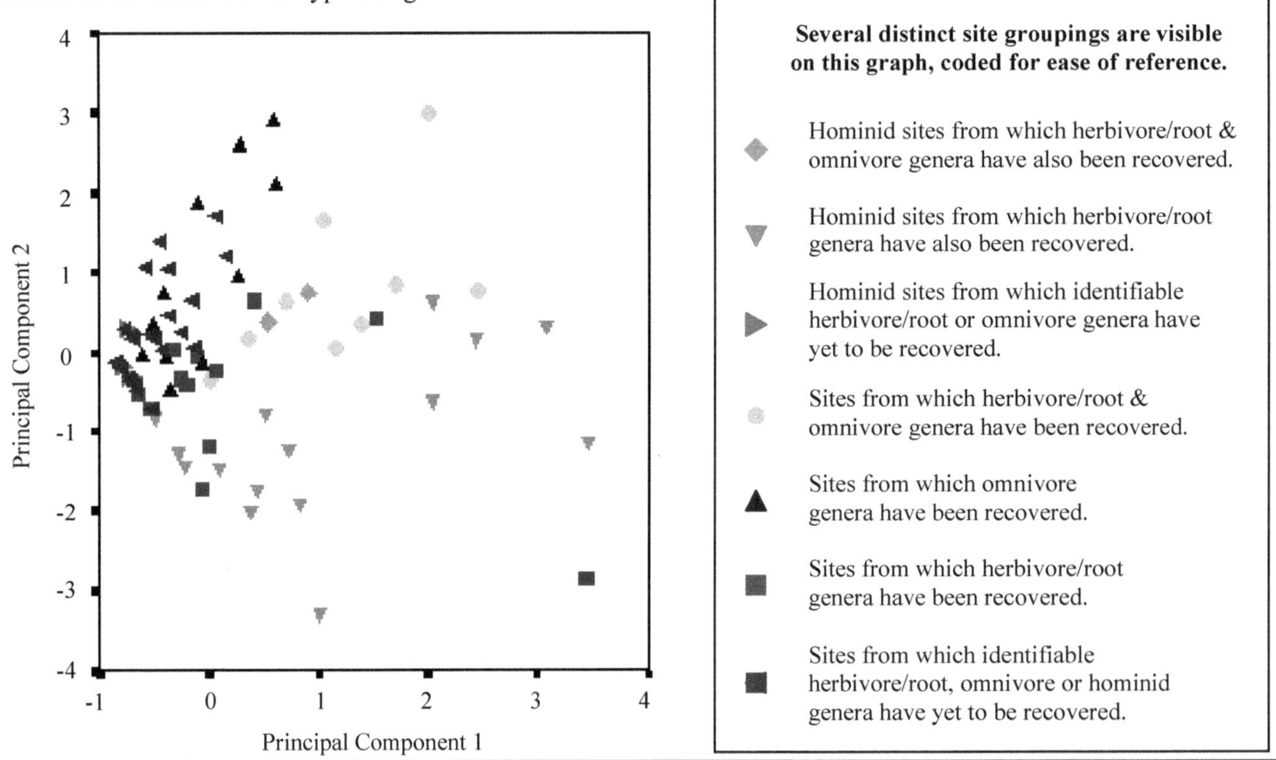

categories. This can be confirmed by comparing Tables 13 and 16. A degree of differentiation has been achieved on component 2, but less than was apparent in the first PCA, and once again the pattern is very similar to that reported in the first analysis, with the carnivore genera separating out from the herbivorous and omnivorous genera; an observation that could have been made from the component matrix in Table 16.

Illustration 18 is a plot of the individual site scores on the first two principal components from the revised diet type groupings, as listed in Table 14. Despite the apparent similarity in component scores between the first and second PCA, Illustration 18 reveals some differences from Illustration 15. There is no clear separation into two sets of data, although some subsets are apparent. This is due to the lower percentage of variance accounted for by the second principal component in the PCA run on the revised diet types compared to that accounted for by the second principal component in the PCA run on the original diet types. Some differentiation of the data on the basis of component 1 is visible, but this has already been attributed to differences in the size of the assemblages under examination and so does not help with interpretation.

Illustration 19 is a plot of the individual site scores on the second and third principal components from the revised diet type groupings, as listed in Table 14. Component 3 is responsible for some additional separation of the data on the basis of the abundance of carnivore/scavenger genera, but the inability of the second principal component to account for much of the variance in the data represents a stumbling block when interpreting the results.

11.3 Discussion

The clear separation of the data into two distinct groupings on the basis of omnivore genera, as shown in Illustration 15, is likely to be somewhat misleading. It reflects an absence of omnivore genera from all sites within Africa that can be attributed to the classificatory scheme used as opposed to a true record of the genera present within Africa during the time period in consideration, which is likely to have included omnivores. Therefore, it would seem inappropriate to base any conclusions on the data analysed using the original diet type categories.

The revised diet type groupings do not result in as clear a separation of the data as was achieved using the original diet types, however, they do represent a more realistic situation than that suggested by the first analysis and important conclusions can still be drawn.

Assemblages from sites from which early hominid remains have been recovered have generally been dominated by herbivorous genera at the expense of carnivorous genera. In fact, hominid remains have been recovered from the vast majority of sites recording a revised second principal component score of less than – 0.6, and hominid remains have also been identified

Illustration 19. Scattergram of Individual Site Scores on the Second and Third Principal Components Based on the Revised Diet Type Categories.

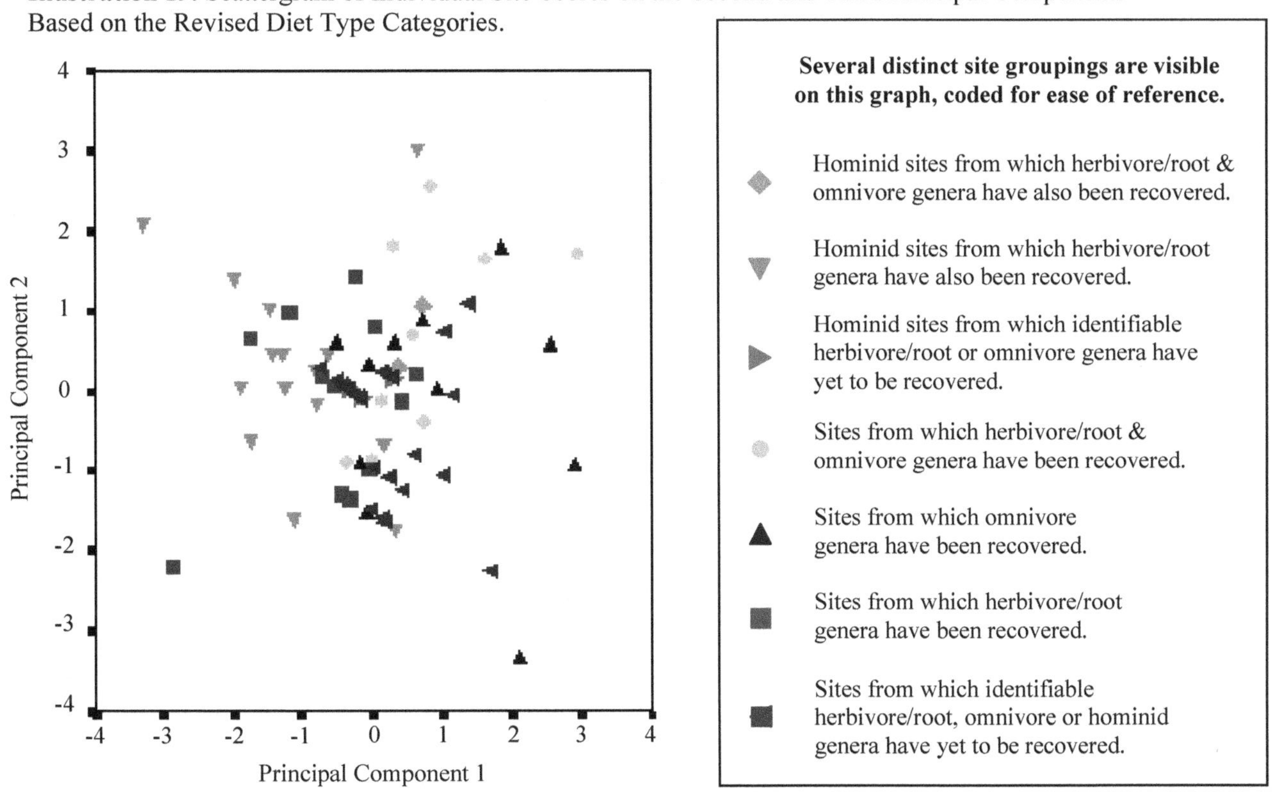

from a few sites recording a revised second principal component score of between 0.74 and – 0.6, whilst at no site with a revised second principal component score of more than 0.74 have hominid remains been identified.

Two of the hominid sites with higher revised second principal component scores are the Eurasian sites of 'Ubeidiya and Dmanisi. Whether this reflects a change in faunal composition outside Africa, a lack of resources forcing different genera to inhabit the same environment or a choice made by these hominids is something that should be examined since it could have implications for the discovery of other hominid sites outside Africa. However, the fact that some African sites, such as Aramis, Olduvai Gorge, Omo and Koobi Fora, also show this pattern of higher revised second principal component scores demonstrates that it is not a phenomenon restricted to Eurasia and so wider explanations may be required. Nevertheless, in all cases the principal component score remains much lower than from sites where carnivore genera dominate the assemblage. This would suggest that, not surprisingly, hominids avoided areas where carnivorous genera were dominant and instead inhabited areas frequented by herbivorous species, which perhaps provided a source of food.

As well as the overall trend for hominid remains to have been identified at sites from which carnivore genera were rare, several subgroups of hominid sites seem to be apparent in the data shown in Illustration 18. However, it must be born in mind that any differentiation between the sites achieved along the horizontal 'x' axis is purely a result of the size of the assemblage analysed. That is to say that points located further to the right on the graph have a larger faunal assemblage than those located to the left. Once this factor has been accounted for it is apparent that most sites from which hominid remains have been identified cluster together reflecting the dominance of herbivore genera described above.

It was hoped that differentiating the herbivore remains on the basis of genera size would shed more light on the apparent correlation between hominid and herbivore remains. All genera within the families Rhinocerotidae, Procyonidae, Gomphotheriidae, Chalicotheriidae, Elephantidae, Deinotheridae, Hippopotamidae and Mammutidae were therefore classified as large herbivores, with the remaining herbivores categorised as small/medium. This alteration did not impact on the overall conclusions obtained from the PCA analysis on the revised diet type categories, but did allow further interpretation of the results.

The number of genera comprising the large herbivore category was considerably smaller than the number of genera in the small/medium herbivore category, which is likely to be as a result of their lower density in the landscape. The only hominid site from which a high percentage of large herbivore genera have been recovered is Fejej, Ethiopia, but few herbivore genera in total were recovered and so this result is insignificant. On the whole, high percentages of large herbivore genera do not appear to be correlated with hominid sites; that is most hominid remains have been found in association with high percentages of small/medium herbivore genera and low or medium percentages of large herbivore genera. The percentage of large herbivore genera is varied across all sites from which hominid remains have not been recovered, although they are most prevalent at sites from which herbivore/root, omnivore and hominid remains are absent and seem to avoid sites with high numbers of carnivore genera. This shows that assemblages from which hominid remains have been recovered tend to be dominated by small/medium herbivores rather than large herbivores. This is not entirely unsurprising given that small/medium herbivores and hominids would have been more likely to suffer from predation than larger herbivores and so were likely to find similar environments the 'safest' for survival. Also, as suggested above, small/medium herbivores may have provided a source of food for hominids who were, as yet, incapable of attacking larger animals.

Two sites from which hominid remains have not, as yet, been identified are Chandigarh, in the Upper Siwaliks, and Hadar, during the Middle Pleistocene period. From the information depicted in Illustration 18, this would appear to be an anomalous result since these sites group closely with a number of African sites from which hominid remains have been identified. This does indicate that additional work at these localities may reveal evidence for hominid activity.

There is less of a clear-cut separation when looking at sites from which hominid remains have been identified and comparing them to sites from which several carnivore/scavenger genera have been recovered, as is shown in Illustration 19. Whilst a large number of hominid sites have a faunal assemblage dominated by genera other than carnivore/scavenger, hominid remains have been identified from several African sites (Omo (3.0 – 1.9 Ma), Olduvai Gorge (1.7 – 0.8 Ma), Koobi Fora and the Chemeron Formation) where the number of carnivore/scavenger genera is relatively high, although the number of pure carnivore genera remains low. This does seem to imply that the environmental requirements, and perhaps food requirements, of carnivore/scavenger genera and hominids in this part of Africa overlapped, with the more specialist carnivore genera inhabiting different areas.

Once again the sites of Chandigarh and Hadar, during the Middle Pleistocene period group with sites from which hominid remains have been identified, further suggesting the success additional work at these localities could bring.

In general, faunal sites from Asia and Europe tend to have more carnivorous genera represented in their assemblages whilst African assemblages are dominated by herbivorous genera. There is a more even distribution in the number of carnivore/ scavenger genera found in assemblages from Africa, Europe and Asia, but once again African sites are at the lower end of the scale. This does imply that there is a distinct difference between faunal assemblages collected from Eurasia and Africa, and this may go some way towards explaining why hominid remains remain unidentified from most sites within Eurasia. The reasons underlying the distinct differences between the faunal assemblages from African and Eurasian sites that has been revealed by the PCA represents an intriguing situation. Whether this difference between the assemblages is a symptom of the restricted amount of excavation that has been carried out over Eurasia, or is a true reflection of the faunal composition during this time period can only be determined once additional work over the area has been undertaken.

11.4 Conclusions

The Principal Components Analysis, as expected, has not revealed any potential new areas in which a search for hominid activity may prove fruitful. Nevertheless, it has helped to explain why hominid remains remain unidentified over the vast majority of the huge area of Eurasia during the Late Pliocene and Early Pleistocene and has suggested some possible new lines of research. The faunal assemblages from most of the Eurasian sites already excavated do not seem to suggest that the immediate environment surrounding the site would have been suitable for early hominids on the basis of comparison with African hominid sites. However, the analysis does seem to suggest that additional work at the site of Chandigarh and the Middle Pleistocene sediments at Hadar may reveal the presence of hominids.

In terms of interpreting the earliest colonisation of Eurasia, the analysis has shown that more work needs to be conducted in order to understand whether the faunal communities of Eurasia were dramatically different from those in Africa and if this would have had an effect on the migration of hominids. The presence of hominids at two sites in Eurasia, Dmanisi and 'Ubeidiya, where there would appear to be slightly increased levels of carnivore and carnivore/scavenger genera when compared to hominid sites in Africa, does imply that this was not a deterrent for hominids when colonising new areas. Also, the fact that these Eurasian sites group quite closely with other faunal sites within Eurasia, particularly in China, suggests that hominids could be identified at other localities.

Chapter 12

Conclusion

Both the GIS simulation model and the statistical analysis of the data collated in this research show that substantial progress in understanding early hominid migrations during the Late Pliocene and Early Pleistocene can be made through interdisciplinary research. An investigation of the many different fields encompassed by this research illustrates the vast array of knowledge that is required in order to comprehend fully not only the factors that would have affected Eurasian mammal migration routes in the Plio-Pleistocene, but also the problems faced by modern researchers in acquiring suitable datasets. It is only when these two strands have been successfully identified and acknowledged that a true appreciation of early hominid migration can be attained.

The Arc-SDM model and the statistical analysis represent inductive, empirical investigations, since they have their basis in observed patterns in data. However, as was indicated in Section 3, no model is ever 100% inductive as a certain amount of deductive reasoning will always be involved. In this instance, the initial hypotheses, based on deductive reasoning, were that a combination of environmental factors would have determined when hominid migration could have taken place and that modern-day politics would have had a dramatic effect on the general understanding of migration pathways. Whilst only the first of these hypotheses could be tested directly through the development of the two models, the lack of data and subsequent inconclusive model results for Central Asia provided indirect confirmation of the second hypothesis. The accuracy of the first hypothesis was established with the successful identification of areas to which early hominids either migrated or could have migrated. This shows that in order to achieve a successful inductive model of early hominid migration, a certain amount of pre-existing knowledge was required in order to identify correctly the variables that would have affected the migration pathways and so would need to be included in any model.

One of the most challenging aspects of this research was the lack of data from substantial parts of the research area. World political history of the 20th Century is mostly able to explain why some regions of the research area were short of data whilst others were much richer. Other issues were involved, however, and they included religious extremism, anti-Western sentiment, low GNP, extremely poor standards of education and geological/geographical factors that resulted in a lack of preservation of Plio-Plesitocene fluvial deposits.

The variables that were thought likely to have influenced hominid migrations and analysed by the two different techniques did not include any socially orientated variables. This is because in the Plio-Pleistocene it is unclear what social behaviours would have been in place in a hominid group and to what extent they would have limited the behaviour of the group in terms of where and when any migration movements could have occurred. It is unlikely that this whole sector of behaviour-affecting variables could be excluded from migration models of more recent hominids, such as *Homo sapiens*. However, the likelihood that social variables would have had a serious impact on migration during the time period under consideration is small. Furthermore, since the scale at which this model operates is so coarse, the effect of any social variables, if they had existed, is unlikely to be discernible.

Once the variables that were to be included in the models had been defined, it was necessary to procure the data, as shown in Illustration 20. The use of palaeovegetation data, primarily from the PRISM2 project (Dowsett et al., 1999), added an enormous strength to the model conclusions, as very different results would have been obtained if modern vegetation data had been used. This is shown in Illustration 4, where the amount of vegetation change between the Pliocene and the modern-day is demonstrated. Surprisingly, many archaeological GIS models, which have been developed for a variety of reasons, have utilised modern-day environmental data as the basis for interpretations (see papers within Lock, 2000; Wescott & Brandon 2000). This assumes that environmental conditions have remained stable, something that may be feasible for models dealing with more recent time periods, but would have been hugely incorrect for the Plio-Pleistocene.

Although the GIS model has not been as comprehensively tested as would be ideal, it has proved to be accurate within the current limitations of available data. Also, further checking of the conclusions can be achieved by conducting fieldwork in the areas identified as high probability and by adding data to the model as and when it becomes available. Nevertheless, the internal checks that were put in place by conducting statistical analysis to independently confirm the results of the GIS model and the in-depth investigation of the palaeogeography assisted with the acceptance of the GIS results.

The overriding result that was visible in every area investigated was that migration out of Africa would have been 'easier' in the Late Pliocene than in the Early Pleistocene, as was suggested by Turner (1999). The onset of glaciation was responsible for the change in climate from one of relatively high humidity to one of increasing aridity, and aridity had generally taken hold by the Pleistocene and had altered the prevailing vegetation. The GIS model, pollen and faunal analysis all concurred

Illustration 20. Step-By-Step Development of the Models to Simulate Early Hominid Migration. Adapted from Pereira & Itami, 1991:1477

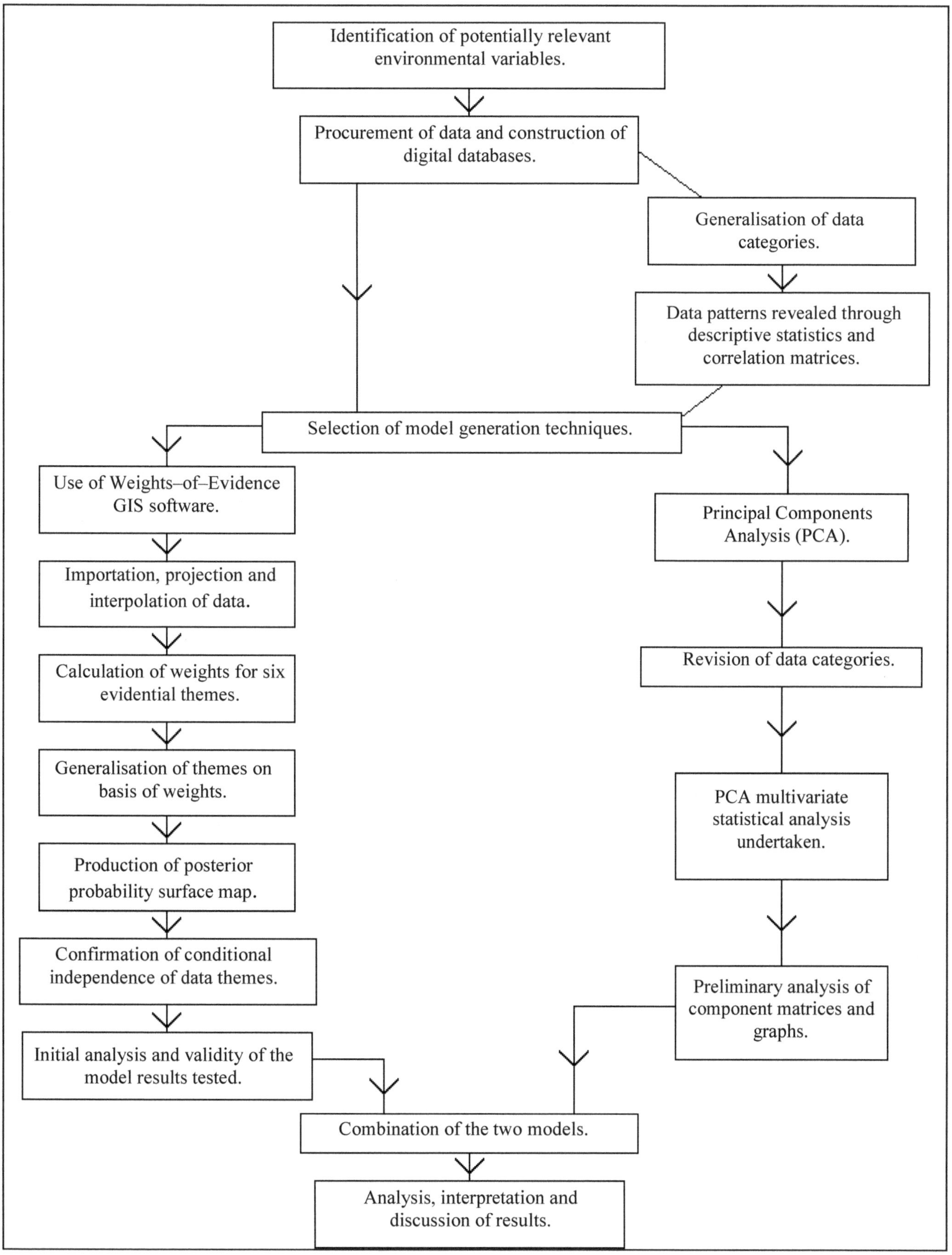

that the Black Sea area probably represented a refugium for at least part of the time period being considered.

The Principal Components Analysis revealed some interesting patterns relating to the differences in faunal assemblage structure between Africa and Eurasia that will require additional work in order to interpret fully. The faunal assemblages from 'Ubeidiya and Dmanisiwere slightly different from many, but not all, of the other hominid sites within East Africa in that there was a higher number of carnivore/ scavenger genera present. In general, however, the results would appear to indicate that hominids were most commonly associated with herbivores, particularly those of a small or medium size, than with carnivores. Several possible conclusions can be drawn from these results. Firstly, the carnivore/scavenger guild may have been much larger in Eurasia than in Africa making it more likely that carnivore/scavenger remains would be found in association with hominid fossils. Hominids and carnivore/scavengers may have preyed on small and medium sized herbivores, resulting in a co-occurrence of their remains in fossil assemblages. It may also be that early hominids had to be able to co-exist and compete with carnivore/scavenger genera before they were successfully able to migrate from the African continent.

The areas identified by the PCA as possible locations for early hominids – Hadar, Chandigarh and some sites in China – were mostly confirmed by the GIS predictive model. The GIS model did not predict Chandigarh as being located in an area with a high probability for hominid sites, and there was a reasonable level of confidence in the result. The Chinese sites of Bijiashan, Houhecun and Yuanmou and the North China area and the North Asian area were closely correlated with 'Ubeidiya and Dmanisi in the PCA, and all except Houhecun were identified by the predictive model as being likely places from which hominid fossils could be recovered. The result from Houhecun, however, was not expressed with much confidence, whilst the confidence levels associated with the other sites were all quite high. The fact that the two different techniques, carried out on the same data, achieved comparable results is encouraging and firmly indicates that these areas would benefit from additional research.

All in all, this research represents a successful first attempt at generating a site predictive model to determine early hominid migration patterns. Explanations for the lack of data from large areas have been provided and new hypotheses relating to the timing and routing of early hominid migrations have been suggested. The required next steps are also clearly defined. As it becomes available, new faunal and floral data should be added to the model database, particularly from sites within Central Asia, and details of any further hominid discoveries should also be included. The most rigorous testing of the model would then be to conduct fieldwork in the areas highlighted as being of great potential for hominid migration and the information obtained subsequently fed back into the model database so that any necessary refinements can be made. In this way the model would continue to represent the most up-to-date findings.

Appendix 1

The Taxonomic Classification and Dietary Requirements of All Mammals Used in the Statistical Analysis and to Generate the Predictive Site Model

Order	Family	Tribe	Genus	Diet
Perissodactyla	Rhinocerotidae		Aceratherium	Herbivore
Carnivora	Felidae		Acinonyx	Carnivore
Artiodactyla	Bovidae	Aepycerotini	Aepyceros	Herbivore
Artiodactyla	Bovidae	Aepycerotini	Aepycerotini	Herbivore
Artiodactyla	Suidae		Afrochoerus	Herbivore/Roots
Carnivora	Ursidae		Agriotherium	Carnivore
Carnivora	Procyonidae		Ailuropoda	Herbivore
Carnivora	Procyonidae		Ailurus	Herbivore
Artiodactyla	Bovidae	Alcelaphini	Alcelaphini	Herbivore
Artiodactyla	Bovidae	Alcelaphini	Alcelaphus	Herbivore
Artiodactyla	Cervidae		Alces	Herbivore
Proboscidea	Gomphotheriidae		Anancus	Herbivore
Perissodactyla	Chalicotheridae		Ancylotherium	Herbivore
Artiodactyla	Bovidae	Antelopini	Antelopini	Herbivore
Artiodactyla	Bovidae	Antelopini	Antidoras	Herbivore
Artiodactyla	Bovidae	Antelopini	Antidorcas	Herbivore
Artiodactyla	Bovidae	Antilopini	Antilope	Herbivore
Artiodactyla	Bovidae	Antilopini	Antilopini	Herbivore
Artiodactyla	Bovidae	Antilopini	Antilospira	Herbivore
Proboscidea	Elephantidae		Archidiskondon	Herbivore
Artiodactyla	Cervidae		Arvernoceros	Herbivore
Primates	Hominidae		Australopithecus	Omnivore/Scavenger
Artiodactyla	Cervidae		Axis	Herbivore
Artiodactyla	Bovidae	Alcelaphini	Beatragus	Herbivore
Artiodactyla	Bovidae	Bovini	Bibos	Herbivore
Artiodactyla	Bovidae	Bovini	Bison	Herbivore
Artiodactyla	Bovidae	Bovini	Bos	Herbivore
Artiodactyla	Bovidae		Bovidae	Herbivore
Artiodactyla	Bovidae	Bovini	Bovini	Herbivore
Perissodactyla	Rhinocerotidae		Brachypotherium	Herbivore
Artiodactyla	Bovidae	Bovini	Bubalus	Herbivore
Artiodactyla	Bovidae	Ovibovini	Budorcas	Herbivore
Artiodactyla	Camelidae		Camelidae	Herbivore
Artiodactyla	Camelidae		Camelus	Herbivore
Carnivora	Canidae		Canidae	Carnivore
Carnivora	Canidae		Canis	Carnivore
Artiodactyla	Bovidae	Caprini	Capra	Herbivore
Artiodactyla	Cervidae		Capreolus	Herbivore
Artiodactyla	Bovidae	Rupicapra	Capricornis	Herbivore
Artiodactyla	Caprinae		Caprinae	Herbivore
Carnivora			Carnivora	Carnivore
Artiodactyla	Bovidae	Cephalophini	Cephalophini	Herbivore
Artiodactyla	Bovidae	Cephalophini	Cephalophus	Herbivore
Perissodactyla	Rhinocerotidae		Ceratotherium	Herbivore
Primates	Cercopithecinae	Papionini	Cercocebus	Herbivore
Primates	Cercopithecidae		Cercopithecidae	Herbivore
Primates	Cercopithecidae		Cercopithecinae	Herbivore
Primates	Colobinae		Cercopithecoides	Herbivore
Primates	Cercopithecidae	Papionini	Cercopithecus	Herbivore
Artiodactyla	Cervidae		Cervalces	Herbivore

Order	Family	Tribe	Genus	Diet
Artiodactyla	Cervidae		Cervidae	Herbivore
Artiodactyla	Cervidae		Cervocerus	Herbivore
Artiodactyla	Cervidae		Cervus	Herbivore
Perissodactyla	Chalicotheriidae		Chalicotherium	Herbivore
Carnivora	Hyaenidae		Chasmaporthetes	Carnivore/Scavenger
Perissodactyla	Rhinocerotidae		Chilotherium	Herbivore
Perissodactyla	Chalicotheriidae		Circotherium	Herbivore
Perissodactyla	Rhinocerotidae		Coelodonta	Herbivore
Primates	Colobidae		Colobinae	Herbivore
Primates	Colobidae		Colobus	Herbivore
Artiodactyla	Bovidae	Alcelaphini	Connochaetes	Herbivore
Carnivora	Hyaenidae		Crocuta	Carnivore/Scavenger
Artiodactyla	Cervidae		Croizetoceros	Herbivore
Carnivora	Canidae		Cuon	Carnivore
Carnivora	Felidae		Cynailurus	Carnivore
Artiodactyla	Cervidae		Dama	Herbivore
Artiodactyla	Bovidae	Alcelaphini	Damaliscus	Herbivore
Artiodactyla	Bovidae	Alcelaphini	Damelops	Herbivore
Artiodactyla	Cervidae		Deer	Herbivore
Proboscidea	Deinotheridae		Deinotherium	Herbivore
Perissodactyla	Rhinocerotidae		Dicerorhinus	Herbivore
Perissodactyla	Rhinocerotidae		Diceros	Herbivore
Artiodactyla	Suidae		Dicoryphochoerus	Herbivore/Roots
Artiodactyla	Suidae		Dicoryphoerus	Herbivore/Roots
Carnivora	Felidae		Dinofelis	Carnivore
Artiodactyla	Bovidae		Dmanisibos	Herbivore
Artiodactyla	Tragulidae		Dorcabune	Herbivore
Artiodactyla	Suidae		Ectopotamochoerus	Herbivore/Roots
Artiodactyla	Cervidae		Elaphodus	Herbivore
Artiodactyla	Cervidae		Elaphurus	Herbivore
Perissodactyla	Rhinocerotidae		Elasmotherium	Herbivore
Probiscidea	Elephantidae		Elephantidae	Herbivore
Proboscidae	Elephantidae		Elephas	Herbivore
Artiodactyla	Bovidae		Eobison	Herbivore
Artiodactyla	Cervidae		Eostyloceros	Herbivore
Artiodactyla	Bovidae	Boselaphini	Eotragus	Herbivore
Carnivora	Felidae		Epimachairodus	Carnivore
Perissodactyla	Equidae		Equidae	Herbivore
Perissodactyla	Equidae		Equus	Herbivore
Artiodactyla	Cervidae		Eucladoceros	Herbivore
Artiodactyla	Cervidae		Euctenoceros	Herbivore
Carnivora	Hyaenidae		Euryboas	Carnivore/Scavenger
Carnivora	Felidae		Felidae	Carnivore
Carnivora	Felidae		Felinae	Carnivore
Carnivora	Felidae		Felis	Carnivore
Artiodactyla	Bovidae		Gallogoral	Herbivore
Artiodactyla	Bovidae	Antelopini	Gazella	Herbivore
Artiodactyla	Bovidae		Gazellini	Herbivore
Artiodactyla	Bovidae	Antilopini	Gazellospira	Herbivore

Order	Family	Tribe	Genus	Diet
Primates	Pongidae		Gigantopithecus	Herbivore
Artiodactyla	Giraffidae		Giraffa	Herbivore
Artiodactyla	Giraffidae		Giraffidae	Herbivore
Proboscidea	Gomphotheriidae		Gomphotherium	Herbivore
Artiodactyla	Bovidae	Bovini	Gorgon	Herbivore
Artiodactyla	Bovidae	Bovini	Hemibos	Herbivore
Artiodactyla	Bovidae	Caprini	Hemitragus	Herbivore
Artiodactyla	Bovidae		Hesperoceros	Herbivore
Artiodactyla	Hippopotamidae		Hexaprotodon	Herbivore
Perissodactyla	Equidae		Hipparion	Herbivore
Perissodactyla	Equidae		Hipparioninae	Herbivore
Artiodactyla	Hippopotimidae		Hippopotamidae	Herbivore
Artiodactyla	Hippopotimidae		Hippopotamus	Herbivore
Artiodactyla	Bovidae	Hippotragini	Hippotragini	Herbivore
Artiodactyla	Bovidae	Hippotragini	Hippotragus	Herbivore
Primates	Hominidae		Hominidae	Omnivore/Scavenger
Primates	Hominidae		Homo	Omnivore/Scavenger
Carnivora	Felidae		Homotherium	Carnivore
Carnivora	Hyaenidae		Hyaena	Carnivore/Scavenger
Carnivora	Hyaenidae		Hyaenea	Carnivore/Scavenger
Carnivora	Hyaenidae		Hyaenidae	Carnivore/Scavenger
Primates	Pongidae		Hylobates	Herbivore
Artiodactyla	Suidae		Hylochoerus	Herbivore/Roots
Carnivora	Hyaenidae		Ictitherium	Carnivore/Scavenger
Artiodactyla	Bovidae	Reduncini	Kobus	Herbivore
Artiodactyla	Suidae		Kolpochoerus	Herbivore/Roots
Carnivora	Hyaenidae		Leecyaena	Carnivore/Scavenger
Artiodactyla	Bovidae	Bovini	Leptobos	Herbivore
Artiodactyla	Giraffidae		Libytherium	Herbivore
Proboscidea	Elephantidae		Loxodonta	Herbivore
Carnivora	Canidae		Lycaeon	Carnivore
Carnivora	Canidae		Lycaon	Carnivore
Carnivora	Felidae		Lynx	Carnivore
Primates	Cercopithecidae		Macaca	Herbivore
Artiodactyla	Giraffidae		Macedonitherium	Herbivore
Carnivora	Felidae		Machairodontinae	Carnivore
Carnivora	Felidae		Machairodus	Carnivore
Artiodactyla	Bovidae	Neotragini	Madoqua	Herbivore
Proboscidea	Mammutidae		Mammut	Herbivore
Proboscidea	Elephantidae		Mammuthus	Herbivore
Proboscidea	Mammutidae		Mastodon	Herbivore
Artiodactyla	Cervidae		Megaceros	Herbivore
Artiodactyla	Cervidae		Megaloceros	Herbivore
Artiodactyla	Bovidae	Alcelaphini	Megalotragus	Herbivore
Artiodactyla	Bovidae	Ovibovini	Megalovis	Herbivore
Primates	Cercopithecidae		Megamacaca	Herbivore
Carnivora	Felidae		Megantereon	Carnivore
Perissodactyla	Tapiridae		Megatapirus	Herbivore
Artiodactyla	Bovidae	Reduncini	Menelikia	Herbivore
Artiodactyla	Anthracotheriidae		Merycopotamus	Herbivore/Roots

Order	Family	Tribe	Genus	Diet
Artiodactyla	Suidae		Mesochoerus	Herbivore/Roots
Artiodactyla	Cervidae		Metacervulus	Herbivore
Carnivora	Felidae		Metailurus	Carnivore
Artiodactyla	Suidae		Metridiochoerus	Herbivore/Roots
Artiodactyla	Cervidae		Muntiacus	Herbivore
Artiodactyla	Bovidae	Neotragini	Neotragini	Herbivore
Perissodactyla	Chalicotheriidae		Nesiortherium	Herbivore
Perissodactyla	Chalicotheriidae		Nestoritherium	Herbivore
Artiodactyla	Suidae		Notochoerus	Herbivore/Roots
Artiodactyla	Bovidae	Alcelaphini	Numidocapra	Herbivore
Artiodactyla	Suidae		Nyanzachoerus	Herbivore/Roots
Carnivora	Canidae		Nyctereutes	Carnivore
Artiodactyla	Giraffidae		Okapia	Herbivore
Artiodactyla	Bovidae	Alcelaphini	Oreonagor	Herbivore
Artiodactyla	Bovidae	Neotragini	Oreotragus	Herbivore
Artiodactyla	Cervidae		Orthogonocerus	Herbivore
Artiodactyla	Suidae		Orthostonyx	Herbivore/Roots
Artiodactyla	Bovidae	Hippotragini	Oryx	Herbivore
Carnivora	Canidae		Otocyon	Carnivore
Artiodactyla	Bovidae	Neotragini	Ourebia	Herbivore
Artiodactyla	Bovidae	Ovibovini	Ovibos	Herbivore
Artiodactyla	Bovidae	Caprini	Ovis	Herbivore
Carnivora	Hyaenidae		Pachycrocuta	Carnivore/Scavenger
Proboscidea	Elephantidae		Palaeoloxoden	Herbivore
Artiodactyla	Giraffidae		Paleotraginae	Herbivore
Carnivora	Felidae		Panthera	Carnivore
Primates	Cercopithecini	Papionini	Papio	Herbivore
Primates	Cercopithecini	Papionini	Papionini	Herbivore
Artiodactyla	Camelidae		Paracamelus	Herbivore
Artiodactyla	Cervidae		Paracervulus	Herbivore
Primates	Colobinae		Paracolobus	Herbivore
Primates	Colobinae		Parapapio	Herbivore
Artiodactyla	Bovidae	Alcelaphini	Parmularis	Herbivore
Artiodactyla	Bovidae	Alcelaphini	Parmularius	Herbivore
Artiodactyla	Bovidae	Bovini	Pelorovis	Herbivore
Artiodactyla	Bovidae	Bovini	Pelovoris	Herbivore
Proboscidea	Gomphotheriidae		Pentalophodon	Herbivore
Artiodactyla	Suidae		Phacochoerus	Herbivore/Roots
Artiodactyla	Bovidae	Antelopini	Phenacotragus	Herbivore
Artiodactyla	Bovidae		Pliotragus	Herbivore
Primates	Pongidae		Pongo	Herbivore
Artiodactyla	Bovidae		Pontoceros	Herbivore
Artiodactyla	Suidae		Potamochoerus	Herbivore/Roots
Artiodactyla	Suidae		Potamochoroides	Herbivore/Roots
Artiodactyla	Cervidae		Praemegaceros	Herbivore
Artiodactyla	Bovidae	Ovibovini	Praeovibos	Herbivore
Proboscidea	Elephantidae		Primelephas	Herbivore
Artiodactyla	Bovidae	Bovini	Proamphibos	Herbivore
Proboscidea			Probiscoidea	Herbivore

Order	Family	Tribe	Genus	Diet
Perissodactyla	Equidae		Proboscidipparion	Herbivore
Artiodactyla	Bovidae	Caprini	Procamptoceras	Herbivore
Artiodactyla	Cervidae		Procapreolus	Herbivore
Primates	Cercopithecidae		Procynoce	Herbivore
Artiodactyla	Suidae		Promesochoerus	Herbivore/Roots
Artiodactyla	Suidae		Pronotochoerus	Herbivore/Roots
Artiodactyla	Cervidae		Pseudaxis	Herbivore
Artiodactyla	Bovidae	Caprini	Pultiphagonides	Herbivore
Artiodactyla	Bovidae	Alcelaphini	Rabaticeras	Herbivore
Artiodactyla	Cervidae		Rangifer	Herbivore
Artiodactyla	Bovidae	Neotragini	Raphicerus	Herbivore
Artiodactyla	Bovidae	Reduncini	Redunca	Herbivore
Artiodactyla	Bovidae	Reduncini	Reduncini	Herbivore
Perissodactyla	Rhinocerotidae		Rhinoceros	Herbivore
Perissodactyla	Rhinocerotidae		Rhinocerotidae	Herbivore
Artiodactyla	Cervidae		Rucervus	Herbivore
Artiodactyla	Bovidae	Rupicaprini	Rupicapra	Herbivore
Artiodactyla	Cervidae		Rusa	Herbivore
Artiodactyla	Bovidae	Saigini	Saiga	Herbivore
Artiodactyla	Bovidae	Bovini	Simatherium	Herbivore
Proboscidea	Gomphotheriidae		Sinomastodon	Herbivore
Artiodactyla	Suidae		Sivachoerus	Herbivore/Roots
Carnivora			Sivapanthera	Carnivore
Artiodactyla	Giraffidae		Sivatherium	Herbivore
Artiodactyla	Bovidae	Ovibovini	Soergelia	Herbivore
Artiodactyla	Bovidae	Reduncini	Splrocerus	Herbivore
Proboscidea	Elephantidae		Stegodon	Herbivore
Proboscidea	Elephantidae		Stegotetrabelodon	Herbivore
Perissodactyla	Rhinocerotidae		Stephanorhinus	Herbivore
Artiodactyla	Bovidae	Tragelaphini	Strepsiceros	Herbivore
Artiodactyla	Bovidae	Tragelaphini	Strepsicerotini	Herbivore
Artiodactyla	Suidae		Stylochoerus	Herbivore/Roots
Perissodactyla	Equidae		Stylohipparion	Herbivore
Artiodactyla	Suidae		Suidae	Herbivore/Roots
Artiodactyla	Suidae		Sus	Herbivore/Roots
Artiodactyla	Bovidae	Cephalophini	Sylvicapra	Herbivore
Artiodactyla	Bovidae	Bovini	Syncerus	Herbivore
Artiodactyla	Suidae		Tapinochoerus	Herbivore/Roots
Perissodactyla	Tapiridae		Tapirus	Herbivore
Artiodactyla	Bovidae	Tragelaphini	Taurotragus	Herbivore
Proboscidea	Gomphotheriidae		Teiralophodon	Herbivore
Proboscidea	Mammutidae		Tetralophodon	Herbivore
Artiodactyla	Bovidae	Reduncini	Thalerocerus	Herbivore
Primates	Cercopithecidae		Theropithecus	Herbivore
Carnivora	Canidae		Thos	Carnivore
Carnivora			Torolutra	Carnivore
Artiodactyla	Bovidae	Tragelaphini	Tragelaphini	Herbivore
Artiodactyla	Bovidae	Tragelaphini	Tragelaphus	Herbivore
Artiodactyla	Bovidae	Alcelaphini	Twisted horn antelope	Herbivore
Artiodactyla	Bovidae	Bovini	Ugandax	Herbivore

Order	Family	Tribe	Genus	Diet
Carnivora	Ursidae		Ursidae	Omnivore
Carnivora	Ursidae		Ursus	Omnivore
Carnivora			Vishnuictis	Carnivore
Artiodactyla	Giraffidae		Vishnutherium	Herbivore
Carnivora	Canidae		Vulpes	Carnivore
Artiodactyla	Bovidae	Alcelaphini	Xenocephalus	Herbivore
Carnivora	Canidae		Xenocyon	Carnivore

For classificatory ease and to determine with what species hominids would have competed for food, all primates, with the exception of hominids, have been classed as herbivores.

Data obtained from: Bar - Yosef, 1995; Beden, 1983; Black, 1978; Butler, 1978; Churcher, 1978; Coppen and Howell, 1976; Gentry, 1976; Gentry, 1978; Gentry and Gentry, 1978; Harris, 1976a; Harris, 1976b; Harris, 1976c; Harris, 1983a; Harris, 1983b; Harris, 1983c; Harris, 1991; Hooijer, 1976; Hooijer, 1978; Howell and Petter, 1976; Jablonski and Gu yumin, 1991; Kalb et al., 1982; Kurtén, 1971; Leakey, 1971; Liu Tungsheng, 1991; Maglio, 1975; Qian Fang, 1991; Vrba, 1980; WoldeGabriel et al., 1994; Xue Xiangxi and Zhang Yunxiang, 1991; Zeng Shaohua and Han Defen, 1991.

Appendix 2

Calculated Weights for Topography and New Binary Reclassification Theme.

Height (m above sea level)	Weight	Binary Reclassification
5345	--	Not Predictive
5255	--	Not Predictive
5247	--	Not Predictive
5229	--	Not Predictive
5193	--	Not Predictive
5181	--	Not Predictive
5173	--	Not Predictive
5102	--	Not Predictive
5088	--	Not Predictive
5033	--	Not Predictive
4934	--	Not Predictive
4861	--	Not Predictive
4788	--	Not Predictive
4770	--	Not Predictive
4644	--	Not Predictive
4535	--	Not Predictive
4363	--	Not Predictive
4321	--	Not Predictive
4286	--	Not Predictive
4209	--	Not Predictive
3915	--	Not Predictive
3730	--	Not Predictive
3505	--	Not Predictive
3401	--	Not Predictive
3327	--	Not Predictive
3270	--	Not Predictive
3093	--	Not Predictive
3045	--	Not Predictive
2716	--	Not Predictive
2610	--	Not Predictive
2548	--	Not Predictive

Height (m above sea level)	Weight	Binary Reclassification
2474	--	Not Predictive
2454	--	Not Predictive
2420	--	Not Predictive
2411	--	Not Predictive
2370	--	Not Predictive
2292	--	Not Predictive
2275	--	Not Predictive
2252	--	Not Predictive
2178	--	Not Predictive
2172	--	Not Predictive
2160	--	Not Predictive
2116	--	Not Predictive
2088	--	Not Predictive
2068	--	Not Predictive
2051	--	Not Predictive
2048	--	Not Predictive
2014	--	Not Predictive
1959	--	Not Predictive
1943	--	Not Predictive
1936	--	Not Predictive
1898	--	Not Predictive
1887	--	Not Predictive
1884	--	Not Predictive
1843	--	Not Predictive
1835	--	Not Predictive
1829	--	Not Predictive
1809	--	Not Predictive
1798	--	Not Predictive
1793	-0.0091	Not Predictive
1788	-0.0252	Not Predictive
1778	-0.0410	Not Predictive

Calculated Weights for Topography and New Binary Reclassification Theme Continued.

Height (m above sea level)	Weight	Binary Reclassification
1735	-0.0565	Not Predictive
1722	-0.0718	Not Predictive
1706	-0.0869	Not Predictive
1692	-0.1018	Not Predictive
1688	-0.1164	Not Predictive
1685	-0.1309	Not Predictive
1680	-0.1451	Not Predictive
1664	-0.1591	Not Predictive
1659	-0.1729	Not Predictive
1658	-0.1866	Not Predictive
1644	-0.2000	Not Predictive
1634	-0.2133	Not Predictive
1633	-0.2264	Not Predictive
1617	-0.2393	Not Predictive
1615	-0.2521	Not Predictive
1608	-0.2647	Not Predictive
1601	-0.2772	Not Predictive
1595	-0.2895	Not Predictive
1591	-0.3016	Not Predictive
1551	-0.3255	Not Predictive
1536	-0.3372	Not Predictive
1534	0.3237	Predictive
1501	0.3125	Predictive
1493	0.3014	Predictive
1480	0.2904	Predictive
1479	0.2796	Predictive
1468	0.2688	Predictive
1447	0.2582	Predictive
1432	0.2477	Predictive
1431	0.2373	Predictive
1428	0.2270	Predictive

Height (m above sea level)	Weight	Binary Reclassification
1424	0.2168	Predictive
1415	0.2067	Predictive
1414	0.1967	Predictive
1413	0.1868	Predictive
1411	0.1770	Predictive
1407	0.1673	Predictive
1406	0.1577	Predictive
1405	0.1482	Predictive
1401	0.1388	Predictive
1389	0.1295	Predictive
1383	0.1202	Predictive
1375	0.1110	Predictive
1374	0.1020	Predictive
1370	0.0840	Not Predictive
1358	0.0752	Not Predictive
1354	0.0664	Not Predictive
1350	0.0578	Not Predictive
1311	0.0491	Not Predictive
1300	0.7370	Predictive
1297	0.7285	Predictive
1294	0.7201	Predictive
1290	0.7117	Predictive
1284	0.6952	Predictive
1279	1.0875	Predictive
1277	1.0794	Predictive
1275	1.0715	Predictive
1268	1.0636	Predictive
1267	1.0557	Predictive
1257	1.0479	Predictive
1256	1.0402	Predictive
1255	1.0325	Predictive

Calculated Weights for Topography and New Binary Reclassification Theme Continued.

Height (m above sea level)	Weight	Binary Reclassification
1247	1.0249	Predictive
1235	1.0173	Predictive
1233	1.6230	Predictive
1229	1.6155	Predictive
1227	1.6081	Predictive
1224	1.6007	Predictive
1222	1.5934	Predictive
1217	1.5861	Predictive
1214	1.5789	Predictive
1213	1.5717	Predictive
1212	1.5575	Predictive
1210	1.5505	Predictive
1199	1.5435	Predictive
1197	1.5366	Predictive
1195	1.5297	Predictive
1190	1.5229	Predictive
1187	1.5161	Predictive
1183	1.5094	Predictive
1182	1.5027	Predictive
1179	1.4960	Predictive
1177	1.4894	Predictive
1169	1.4829	Predictive
1168	1.4764	Predictive
1165	1.4699	Predictive
1148	1.4635	Predictive
1147	1.4571	Predictive
1144	1.4507	Predictive
1141	1.4444	Predictive
1138	1.4381	Predictive
1130	1.4319	Predictive
1129	1.4257	Predictive
1124	1.4195	Predictive
1120	1.4134	Predictive
1117	1.4073	Predictive
1106	1.4013	Predictive
1101	1.3953	Predictive
1100	1.3893	Predictive
1098	1.3834	Predictive
1097	1.3775	Predictive
1093	1.3716	Predictive
1091	1.3658	Predictive
1090	1.3600	Predictive
1087	1.3485	Predictive
1082	1.3428	Predictive
1077	1.3371	Predictive
1076	1.3259	Predictive
1073	1.3203	Predictive
1072	1.3147	Predictive
1071	1.3038	Predictive
1069	1.2983	Predictive
1067	1.2929	Predictive
1058	1.2875	Predictive
1052	1.2821	Predictive
1031	1.2715	Predictive
1030	1.2662	Predictive
1027	1.2610	Predictive
1015	1.2557	Predictive
1010	1.2505	Predictive
1006	1.2402	Predictive
1005	1.2351	Predictive
1002	1.2300	Predictive
997	1.2249	Predictive

Calculated Weights for Topography and New Binary Reclassification Theme Continued.

Height (m above sea level)	Weight	Binary Reclassification
993	1.2199	Predictive
991	1.2149	Predictive
988	1.2099	Predictive
985	1.2049	Predictive
983	1.1951	Predictive
982	1.1902	Predictive
981	1.1805	Predictive
980	1.1757	Predictive
978	1.1709	Predictive
968	1.1661	Predictive
965	1.1613	Predictive
962	1.1519	Predictive
959	1.1472	Predictive
958	1.1425	Predictive
957	1.1333	Predictive
956	1.1287	Predictive
955	1.1241	Predictive
952	1.1195	Predictive
944	1.1105	Predictive
943	1.1015	Predictive
940	1.0926	Predictive
934	1.0882	Predictive
930	1.0838	Predictive
929	1.0794	Predictive
924	1.0751	Predictive
922	1.0707	Predictive
921	1.0664	Predictive
920	1.0578	Predictive
916	1.0536	Predictive
907	1.0451	Predictive
903	1.0409	Predictive

Height (m above sea level)	Weight	Binary Reclassification
899	1.0367	Predictive
898	1.0325	Predictive
897	1.0283	Predictive
895	1.0201	Predictive
891	1.0160	Predictive
884	1.0119	Predictive
876	0.9997	Predictive
873	0.9957	Predictive
872	0.9917	Predictive
865	0.9877	Predictive
858	0.9837	Predictive
834	0.9797	Predictive
833	0.9758	Predictive
828	0.9680	Predictive
824	0.9641	Predictive
822	0.9602	Predictive
821	0.9563	Predictive
820	0.9524	Predictive
815	0.9486	Predictive
811	0.9448	Predictive
806	0.9410	Predictive
804	0.9372	Predictive
802	0.9334	Predictive
799	0.9259	Predictive
791	0.9221	Predictive
786	0.9184	Predictive
785	0.9147	Predictive
784	0.9110	Predictive
782	0.9073	Predictive
777	0.9000	Predictive
772	0.8963	Predictive

Calculated Weights for Topography and New Binary Reclassification Theme Continued.

Height (m above sea level)	Weight	Binary Reclassification
769	0.8927	Predictive
768	0.8891	Predictive
767	0.8855	Predictive
764	0.8819	Predictive
763	0.9624	Predictive
760	0.9589	Predictive
757	0.9518	Predictive
755	0.9483	Predictive
752	0.9448	Predictive
750	0.9413	Predictive
746	0.9378	Predictive
744	0.9343	Predictive
739	0.9309	Predictive
736	0.9240	Predictive
730	0.9206	Predictive
723	0.9172	Predictive
721	0.9138	Predictive
720	0.9104	Predictive
718	0.9070	Predictive
715	0.9036	Predictive
712	0.9003	Predictive
709	0.8969	Predictive
707	0.8936	Predictive
702	0.8903	Predictive
699	0.8837	Predictive
695	0.8804	Predictive
693	0.8771	Predictive
692	0.8739	Predictive
688	0.8706	Predictive
687	0.8674	Predictive
686	0.9416	Predictive

Height (m above sea level)	Weight	Binary Reclassification
684	0.9383	Predictive
683	0.9351	Predictive
682	0.9319	Predictive
680	0.9288	Predictive
679	0.9256	Predictive
677	0.9224	Predictive
675	0.9161	Predictive
673	0.9130	Predictive
672	0.9099	Predictive
669	0.9067	Predictive
668	0.9005	Predictive
667	0.8975	Predictive
663	0.8944	Predictive
661	0.8913	Predictive
656	0.8822	Predictive
653	0.8791	Predictive
652	0.9508	Predictive
649	0.9478	Predictive
644	0.9418	Predictive
641	0.9388	Predictive
639	0.9358	Predictive
637	0.9328	Predictive
635	0.9299	Predictive
634	0.9269	Predictive
633	0.9240	Predictive
626	0.9211	Predictive
625	0.9181	Predictive
620	0.9123	Predictive
618	0.9065	Predictive
616	0.9036	Predictive
615	0.8979	Predictive

Calculated Weights for Topography and New Binary Reclassification Theme Continued.

Height (m above sea level)	Weight	Binary Reclassification
606	0.8893	Predictive
605	0.8865	Predictive
598	0.8837	Predictive
597	0.8781	Predictive
594	0.8725	Predictive
589	0.8669	Predictive
588	0.8641	Predictive
585	0.8586	Predictive
581	0.8559	Predictive
580	0.8504	Predictive
578	0.8422	Predictive
577	0.8395	Predictive
576	0.8368	Predictive
571	0.8341	Predictive
568	0.8288	Predictive
566	0.8261	Predictive
565	0.8235	Predictive
560	0.8208	Predictive
558	0.8182	Predictive
557	0.8129	Predictive
556	0.8103	Predictive
555	0.8076	Predictive
553	0.8050	Predictive
550	0.8024	Predictive
549	0.7998	Predictive
548	0.7972	Predictive
545	0.7946	Predictive
542	0.7921	Predictive
536	0.7895	Predictive
535	0.7869	Predictive
534	0.7818	Predictive

Height (m above sea level)	Weight	Binary Reclassification
533	0.7767	Predictive
530	0.7742	Predictive
529	0.7717	Predictive
527	0.7666	Predictive
518	0.7641	Predictive
516	0.7616	Predictive
515	0.7542	Predictive
513	0.7517	Predictive
510	0.7492	Predictive
509	0.7468	Predictive
507	0.7443	Predictive
504	0.7419	Predictive
500	0.7394	Predictive
499	0.7345	Predictive
498	0.7321	Predictive
497	0.7297	Predictive
495	0.7249	Predictive
493	0.7225	Predictive
492	0.7201	Predictive
490	0.7153	Predictive
489	0.7129	Predictive
488	0.7105	Predictive
487	0.7034	Predictive
485	0.7011	Predictive
483	0.6987	Predictive
482	0.6940	Predictive
479	0.6871	Predictive
478	0.6824	Predictive
477	0.6801	Predictive
473	0.6778	Predictive
472	0.6755	Predictive

Calculated Weights for Topography and New Binary Reclassification Theme Continued.

Height (m above sea level)	Weight	Binary Reclassification
471	0.6732	Predictive
470	0.6710	Predictive
467	0.6687	Predictive
464	0.6664	Predictive
463	0.6619	Predictive
459	0.6574	Predictive
458	0.6529	Predictive
457	0.6506	Predictive
456	0.6440	Predictive
455	0.6417	Predictive
454	0.6395	Predictive
451	0.6373	Predictive
447	0.6351	Predictive
443	0.6329	Predictive
442	0.6307	Predictive
441	0.6242	Predictive
436	0.6198	Predictive
435	0.6155	Predictive
434	0.6133	Predictive
432	0.6090	Predictive
430	0.6069	Predictive
429	0.6048	Predictive
428	0.6026	Predictive
425	0.6005	Predictive
424	0.5984	Predictive
422	0.5941	Predictive
421	0.5920	Predictive
420	0.5899	Predictive
417	0.5878	Predictive
416	0.5857	Predictive
414	0.5836	Predictive
412	0.5816	Predictive
411	0.5795	Predictive
410	0.5774	Predictive
409	0.5753	Predictive
408	0.5712	Predictive
406	0.5691	Predictive
405	0.5671	Predictive
401	0.5650	Predictive
400	0.5629	Predictive
397	0.5609	Predictive
396	0.5589	Predictive
395	0.5548	Predictive
394	0.5507	Predictive
393	0.5487	Predictive
390	0.5427	Predictive
389	0.5407	Predictive
388	0.5387	Predictive
387	0.5367	Predictive
385	0.5347	Predictive
384	0.5287	Predictive
381	0.5268	Predictive
379	0.5208	Predictive
378	0.5169	Predictive
376	0.5150	Predictive
375	0.5111	Predictive
374	0.5091	Predictive
369	0.5072	Predictive
368	0.5053	Predictive
366	0.5033	Predictive
362	0.5014	Predictive
361	0.4995	Predictive

Calculated Weights for Topography and New Binary Reclassification Theme Continued.

Height (m above sea level)	Weight	Binary Reclassification
360	0.4976	Predictive
359	0.4956	Predictive
356	0.4937	Predictive
352	0.4918	Predictive
350	0.4899	Predictive
349	0.4861	Predictive
347	0.4804	Predictive
344	0.4786	Predictive
340	0.4767	Predictive
339	0.4729	Predictive
337	0.4711	Predictive
336	0.4673	Predictive
334	0.4618	Predictive
333	0.4599	Predictive
331	0.4562	Predictive
329	0.4544	Predictive
327	0.4507	Predictive
326	0.4489	Predictive
324	0.4471	Predictive
321	0.4452	Predictive
320	0.4434	Predictive
319	0.4416	Predictive
318	0.4380	Predictive
314	0.4344	Predictive
312	0.4326	Predictive
310	0.4308	Predictive
309	0.4290	Predictive
306	0.4272	Predictive
305	0.4254	Predictive
304	0.4930	Predictive
302	0.4894	Predictive

Height (m above sea level)	Weight	Binary Reclassification
300	0.4859	Predictive
299	0.4841	Predictive
298	0.4823	Predictive
297	0.4806	Predictive
294	0.4788	Predictive
291	0.4771	Predictive
288	0.4753	Predictive
287	0.4736	Predictive
285	0.4718	Predictive
284	0.4701	Predictive
282	0.4666	Predictive
281	0.4649	Predictive
280	0.4614	Predictive
277	0.4579	Predictive
276	0.4545	Predictive
275	0.4494	Predictive
273	0.4460	Predictive
271	0.4443	Predictive
270	0.4426	Predictive
268	0.4392	Predictive
267	0.4358	Predictive
265	0.4325	Predictive
264	0.4291	Predictive
263	0.4241	Predictive
261	0.4225	Predictive
260	--	Not Predictive
257	--	Not Predictive
255	--	Not Predictive
253	--	Not Predictive
251	--	Not Predictive
249	--	Not Predictive

Calculated Weights for Topography and New Binary Reclassification Theme Continued.

Height (m above sea level)	Weight	Binary Reclassification
248	--	Not Predictive
245	--	Not Predictive
243	--	Not Predictive
242	--	Not Predictive
241	--	Not Predictive
240	--	Not Predictive
238	--	Not Predictive
235	--	Not Predictive
233	--	Not Predictive
232	--	Not Predictive
231	--	Not Predictive
230	--	Not Predictive
229	--	Not Predictive
227	--	Not Predictive
226	--	Not Predictive
225	--	Not Predictive
224	--	Not Predictive
223	--	Not Predictive
222	--	Not Predictive
219	--	Not Predictive
218	--	Not Predictive
217	--	Not Predictive
216	--	Not Predictive
215	--	Not Predictive
214	--	Not Predictive
212	--	Not Predictive
211	--	Not Predictive
208	--	Not Predictive
207	--	Not Predictive
206	--	Not Predictive
203	--	Not Predictive
201	--	Not Predictive
200	--	Not Predictive
199	--	Not Predictive
197	--	Not Predictive
195	--	Not Predictive
193	--	Not Predictive
192	--	Not Predictive
191	--	Not Predictive
189	--	Not Predictive
188	--	Not Predictive
185	--	Not Predictive
184	--	Not Predictive
183	--	Not Predictive
178	--	Not Predictive
177	--	Not Predictive
176	--	Not Predictive
173	--	Not Predictive
171	--	Not Predictive
169	--	Not Predictive
167	--	Not Predictive
165	--	Not Predictive
161	--	Not Predictive
159	--	Not Predictive
158	--	Not Predictive
157	--	Not Predictive
155	--	Not Predictive
154	--	Not Predictive
152	--	Not Predictive
151	--	Not Predictive
150	--	Not Predictive
148	--	Not Predictive

Calculated Weights for Topography and New Binary Reclassification Theme Continued.

Height (m above sea level)	Weight	Binary Reclassification
147	--	Not Predictive
146	--	Not Predictive
142	--	Not Predictive
138	--	Not Predictive
137	--	Not Predictive
135	--	Not Predictive
134	--	Not Predictive
133	--	Not Predictive
132	--	Not Predictive
131	--	Not Predictive
130	--	Not Predictive
129	--	Not Predictive
124	--	Not Predictive
122	--	Not Predictive
121	--	Not Predictive
120	--	Not Predictive
119	--	Not Predictive
118	--	Not Predictive
117	--	Not Predictive
116	--	Not Predictive
114	--	Not Predictive
113	--	Not Predictive
110	--	Not Predictive
109	--	Not Predictive
108	--	Not Predictive
107	--	Not Predictive
105	--	Not Predictive
102	--	Not Predictive
101	--	Not Predictive
100	--	Not Predictive
99	--	Not Predictive
98	--	Not Predictive
97	--	Not Predictive
96	--	Not Predictive
95	--	Not Predictive
94	--	Not Predictive
93	--	Not Predictive
91	--	Not Predictive
90	--	Not Predictive
89	--	Not Predictive
88	--	Not Predictive
87	--	Not Predictive
86	--	Not Predictive
85	--	Not Predictive
84	--	Not Predictive
77	--	Not Predictive
76	--	Not Predictive
75	--	Not Predictive
74	--	Not Predictive
73	--	Not Predictive
72	--	Not Predictive
71	--	Not Predictive
69	--	Not Predictive
67	--	Not Predictive
66	--	Not Predictive
64	--	Not Predictive
61	--	Not Predictive
59	--	Not Predictive
57	--	Not Predictive
56	--	Not Predictive
55	--	Not Predictive
54	--	Not Predictive

Calculated Weights for Topography and New Binary Reclassification Theme Continued.

Height (m above sea level)	Weight	Binary Reclassification
53	--	Not Predictive
52	--	Not Predictive
51	--	Not Predictive
50	--	Not Predictive
48	--	Not Predictive
47	--	Not Predictive
45	--	Not Predictive
42	--	Not Predictive
41	--	Not Predictive
40	--	Not Predictive
39	--	Not Predictive
38	--	Not Predictive

Height (m above sea level)	Weight	Binary Reclassification
37	--	Not Predictive
36	--	Not Predictive
34	--	Not Predictive
33	--	Not Predictive
32	--	Not Predictive
30	--	Not Predictive
28	--	Not Predictive
27	--	Not Predictive
26	--	Not Predictive
24	--	Not Predictive
0	--	Not Predictive

Calculated Weights for Number of Herbivore Genera and New Binary Reclassification Theme.

Number of Herbivore Genera	Weight	Binary Reclassification
164	4.3527	Predictive
127	3.5450	Predictive
126	3.1041	Predictive
122	2.7991	Predictive
111	3.3004	Predictive
102	3.1041	Predictive
99	2.7991	Predictive
97	2.5658	Predictive
96	2.4668	Predictive
95	2.3767	Predictive
91	2.2177	Predictive
89	2.1468	Predictive
88	2.0806	Predictive
85	2.0185	Predictive
84	1.9048	Predictive
82	1.8525	Predictive
81	1.8027	Predictive
80	1.7554	Predictive
78	1.6254	Predictive
77	1.5472	Predictive
75	1.4748	Predictive
74	1.3750	Predictive
73	1.3439	Predictive
72	1.2559	Predictive
71	1.1002	Predictive
70	1.0532	Predictive

Calculated Weights for Number of Herbivore Genera and New Binary Reclassification Theme Continued.

Number of Herbivore Genera	Weight	Binary Reclassification
69	0.9655	Predictive
68	0.9043	Predictive
67	0.8468	Predictive
66	0.7923	Predictive
65	0.6916	Predictive
64	0.6601	Predictive
63	0.6296	Predictive
62	0.5713	Predictive
61	0.5297	Predictive
60	0.4768	Predictive
59	0.4144	Predictive
58	0.3904	Predictive
57	0.3556	Predictive
56	0.3219	Predictive
55	0.2787	Predictive
54	0.2272	Predictive
53	0.1687	Predictive
52	0.1135	Predictive
51	0.0697	Not Predictive
50	0.0195	Not Predictive
49	-0.0283	Not Predictive
48	-0.0813	Not Predictive
47	-0.1175	Not Predictive
46	-0.1661	Not Predictive
45	-0.2253	Not Predictive
44	-0.2811	Not Predictive
43	-0.2991	Not Predictive
42	-0.3398	Not Predictive
41	-0.3789	Not Predictive
40	-0.4217	Not Predictive
39	-0.4476	Not Predictive
38	-0.4975	Not Predictive
37	-0.1432	Not Predictive
36	-0.1706	Not Predictive
35	-0.1840	Not Predictive
34	-0.2359	Not Predictive
33	-0.2731	Not Predictive
32	-0.3050	Not Predictive
31	-0.3473	Not Predictive
30	-0.3697	Not Predictive
29	-0.4235	Not Predictive
28	-0.4612	Not Predictive
27	-0.4645	Not Predictive
26	-0.5007	Not Predictive
25	-0.0391	Not Predictive
24	-0.0604	Not Predictive
23	0.0697	Not Predictive
22	0.0168	Not Predictive
21	-0.0309	Not Predictive
20	0.0907	Not Predictive

Calculated Weights for Number of Herbivore Genera and New Binary Reclassification Theme Continued.

Number of Herbivore Genera	Weight	Binary Reclassification
19	0.0611	Not Predictive
18	0.0230	Not Predictive
17	0.3420	Predictive
16	0.3286	Predictive
15	0.4067	Predictive
14	0.4577	Predictive
13	0.5111	Predictive
12	0.4709	Predictive
11	0.5162	Predictive
10	0.4879	Predictive
9	0.4658	Predictive
8	0.4406	Predictive
7	0.3989	Predictive
6	0.3837	Predictive
5	0.3638	Predictive
4	0.3410	Predictive
3	0.3172	Predictive
2	0.2863	Predictive
1	0.2652	Predictive
0	0.2187	Predictive
-1	0.1948	Predictive
-2	0.1728	Predictive
-3	0.1513	Predictive
-4	--	Not Predictive
-5	--	Not Predictive
-6	--	Not Predictive
-7	--	Not Predictive
-8	--	Not Predictive
-9	--	Not Predictive
-10	--	Not Predictive
-11	--	Not Predictive
-12	--	Not Predictive
-13	--	Not Predictive
-14	--	Not Predictive
-15	--	Not Predictive
-16	--	Not Predictive
-17	--	Not Predictive
-18	--	Not Predictive
-19	--	Not Predictive
-20	--	Not Predictive
-21	--	Not Predictive
-22	--	Not Predictive
-23	--	Not Predictive
-24	--	Not Predictive
-25	--	Not Predictive
-26	--	Not Predictive
-27	--	Not Predictive
-28	--	Not Predictive
-29	--	Not Predictive
-30	--	Not Predictive

Calculated Weights for Number of Herbivore Genera and New Binary Reclassification Theme Continued.

Number of Herbivore Genera	Weight	Binary Reclassification
-31	--	Not Predictive
-32	--	Not Predictive
-33	--	Not Predictive
-34	--	Not Predictive
-35	--	Not Predictive
-37	--	Not Predictive
-38	--	Not Predictive
-39	--	Not Predictive
-40	--	Not Predictive
-41	--	Not Predictive
-42	--	Not Predictive
-43	--	Not Predictive
-44	--	Not Predictive
-45	--	Not Predictive
-46	--	Not Predictive
-47	--	Not Predictive
-48	--	Not Predictive
-49	--	Not Predictive
-50	--	Not Predictive
-51	--	Not Predictive
-52	--	Not Predictive
-53	--	Not Predictive
-54	--	Not Predictive
-55	--	Not Predictive
-56	--	Not Predictive
-57	--	Not Predictive
-59	--	Not Predictive
-60	--	Not Predictive
-61	--	Not Predictive
-62	--	Not Predictive
-63	--	Not Predictive
-65	--	Not Predictive
-66	--	Not Predictive
-67	--	Not Predictive
-68	--	Not Predictive
-69	--	Not Predictive
-70	--	Not Predictive
-71	--	Not Predictive
-73	--	Not Predictive
-74	--	Not Predictive
-75	--	Not Predictive
-76	--	Not Predictive
-77	--	Not Predictive
-78	--	Not Predictive
-79	--	Not Predictive
-80	--	Not Predictive
-81	--	Not Predictive
-82	--	Not Predictive
-83	--	Not Predictive
-84	--	Not Predictive

Calculated Weights for Number of Herbivore Genera and New Binary Reclassification Theme Continued.

Number of Herbivore Genera	Weight	Binary Reclassification
-85	--	Not Predictive
-86	--	Not Predictive
-87	--	Not Predictive
-88	--	Not Predictive
-89	--	Not Predictive
-90	--	Not Predictive
-91	--	Not Predictive
-92	--	Not Predictive
-93	--	Not Predictive
-95	--	Not Predictive
-96	--	Not Predictive

Calculated Weights for Number of Herbivore/Root Genera and New Binary Reclassification Theme.

Number of Herbivore/Root Genera	Weight	Binary Reclassification
16	--	Not Predictive
15	--	Not Predictive
14	--	Not Predictive
13	--	Not Predictive
12	--	Not Predictive
11	--	Not Predictive
10	--	Not Predictive
9	--	Not Predictive
8	-0.1928	Not Predictive
7	-0.5991	Not Predictive
6	-1.0785	Not Predictive
5	-0.7191	Not Predictive
4	-0.0988	Not Predictive
3	-0.2442	Not Predictive
2	0.0359	Not Predictive
1	0.1879	Predictive
0	--	Not Predictive
-1	--	Not Predictive
-2	--	Not Predictive
-3	--	Not Predictive
-4	--	Not Predictive
-5	--	Not Predictive
-6	--	Not Predictive
-7	--	Not Predictive
-8	--	Not Predictive
-9	--	Not Predictive
-10	--	Not Predictive

Calculated Weights for Number of Carnivore Genera and New Binary Reclassification Theme.

Number of Carnivore Genera	Weight	Binary Reclassification
36	4.3527	Predictive
29	3.1041	Predictive
26	2.7991	Predictive
23	3.3004	Predictive
22	3.1041	Predictive
20	2.9400	Predictive
19	2.5658	Predictive
16	2.3767	Predictive
15	2.2177	Predictive
14	2.0806	Predictive
13	1.9601	Predictive
12	1.7554	Predictive
11	1.4748	Predictive
10	1.2559	Predictive
9	0.9655	Predictive
8	1.2012	Predictive
7	0.7295	Predictive
6	0.1255	Predictive
5	0.3465	Predictive
4	0.2544	Predictive
3	0.0114	Not Predictive
2	0.1814	Predictive
1	0.2858	Predictive
0	--	Not Predictive
-1	--	Not Predictive
-2	--	Not Predictive
-3	--	Not Predictive
-4	--	Not Predictive
-5	--	Not Predictive
-6	--	Not Predictive
-7	--	Not Predictive
-8	--	Not Predictive
-9	--	Not Predictive
-10	--	Not Predictive
-11	--	Not Predictive
-12	--	Not Predictive
-13	--	Not Predictive
-14	--	Not Predictive
-15	--	Not Predictive
-16	--	Not Predictive
-17	--	Not Predictive
-18	--	Not Predictive
-19	--	Not Predictive
-20	--	Not Predictive
-21	--	Not Predictive
-22	--	Not Predictive

Calculated Weights for Number of Carnivore Genera and New Binary Reclassification Theme Continued.

Number of Carnivore Genera	Weight	Binary Reclassification
-23	--	Not Predictive
-24	--	Not Predictive
-25	--	Not Predictive
-26	--	Not Predictive
-27	--	Not Predictive
-28	--	Not Predictive
-29	--	Not Predictive
-30	--	Not Predictive
-31	--	Not Predictive
-32	--	Not Predictive

Calculated Weights for Number of Carnivore/Scavenger Genera and New Binary Reclassification Theme.

Number of Carnivore/Scavenger Genera	Weight	Binary Reclassification
15	4.3527	Predictive
13	2.7991	Predictive
12	1.9601	Predictive
11	1.5855	Predictive
10	1.0084	Predictive
9	1.4404	Predictive
8	1.2844	Predictive
7	1.1245	Predictive
6	0.9655	Predictive
5	0.4768	Predictive
4	-0.0126	Not Predictive
3	-0.3567	Not Predictive
2	-0.8331	Not Predictive
1	0.3829	Predictive
0	--	Not Predictive
-1	--	Not Predictive
-2	--	Not Predictive
-3	--	Not Predictive
-4	--	Not Predictive
-5	--	Not Predictive
-6	--	Not Predictive
-7	--	Not Predictive

Calculated Weights for Vegetation Type and New Binary Reclassification Theme.

Vegetation Type	Weight	Binary Reclassification
Water	--	Not Predictive
Tundra	--	Not Predictive
Coniferous Forest	0.6855	Predictive
Deciduous Forest	-0.9972	Not Predictive
Grassland	0.4167	Predictive
Desert	-0.7498	Not Predictive
Rainforest	--	Not Predictive

Bibliography

Adam, D.P. 1994 Pliocene pollen data set dynamics: Tulelake, California and Lost Chicken Mine, Alaska. In *Pliocene Terrestrial Environments and Data/Model Comparisons.* Open-File Report 94 – 23. Thompson, R.S. (ed.) Washington: U. S. Geological Survey pp 6 – 10.

Adam, D.P., Bradbury, J.P., Rieck, H.J. and Sarna-Wojcicki, A.M. 1990 Environmental changes in the Tulelake basin, Siskiyou and Modoc Counties, California, from 3 to 2 million years before present. *U.S. Geological Survey Bulletin* 1933: 1 – 13.

Adam, D.P., Sarna-Wojcicki, A.M., Rieck, H.J., Bradbury, J.P., Dean, W.E. and Forester, R.M. 1989 Tulelake, California: the last 3 million years. *Palaeogeography, Palaeoclimatology, Palaeoecology* 72: 89 – 103.

Ager, T.A. 1994 Terrestrial palynological and paleobotanical records of Pliocene age from Alaska and Yukon Territory. In *Pliocene Terrestrial Environments and Data/Model Comparisons.* Open-File Report 94 - 23. Thompson, R.S. (ed.) Washington: U. S. Geological Survey pp 2 - 3.

Akhmetiev, M.A. 1991 Flora, vegetation, and climate of Iceland during the Pliocene. Pliocene Climates of the Northern Hemisphere. *U.S. Geological Survey Open-File Report* 91 – 447: 8 – 9.

Akhmetiev, M.A., G. M. Bratoeva, Giterman, R.E., Golubeva, L.V. and Moiseeya, A.I. 1978 Late Cenozoic stratigraphy and flora of Iceland. *Transactions, Academy of Sciences of the USSR* 316.

Al-Farsy, F. 1986 *Saudi Arabia. A Case Study in Development.* London: KPI.

Al-Sayari, S.S. and Zötl, J.G. (eds.) 1978 *Quaternary Period in Saudi Arabia. Sedimentological, Hydrogeological, Hydrochemical, Geomorphological and Climatological Investigations in Central and Eastern Saudi Arabia.* Wien & New York: Springer-Verlag.

Al-Sulaimi, J.S. and Pitty, A.F. 1995 Origin and depositional model of Wadi Al-Batin and its associated alluvial fan, Saudi Arabia and Kuwait. *Sedimentary Geology* 97: 203 – 229.

Alavi, M. 1994 Tectonics of the Zagros orogenic belt of Iran – new data and interpretations. *Tectonophysics* 229 (3 – 4): 211 – 238.

Alberdi, M.T., Prado, J.L. and Ortiz-Jaureguizar, E. 1995 Patterns of body size changes in fossil and living Equini (Perissodactyla). *Biological Journal of the Linnean Society* 54: 349 – 370.

Allen, K.M.S., Green, S.W. and Zubrow, E.B.W. 1990 Interpreting space. In *Interpreting Space: GIS and Archaeology.* Allen, K.M.S., Green, S.W. and Zubrow, E.B.W. (eds.) London: Taylor and Francis pp 383 – 386.

An Zhisheng, Kutzbach, J.E., Prell, W.L. and Porter, S.C. 2001 Evolution of Asian monsoons and phased uplift of the Himalaya-Tibetan plateau since Late Miocene times. *Nature* 411: 62 – 66.

Andersson, C. 1997 Transfer function vs. modern analog technique for estimating Pliocene sea surface temperatures based on planktic foraminiferal data, western equatorial Pacific Ocean. *Journal of Foraminiferal Research* 27: 123 – 132.

Andrews, P. 1992 The basis for taphonomic research on vertebrate fossils. In *Conferencias de la Reunión de Tafonomia y Fosilización.* López, S.F. (ed.) Madrid: Editorial Complutense pp 33 – 43.

Andrews, P. 1996 Palaeoecology and hominoid palaeoenvironments. *Biological Reviews of the Cambridge Philosophical Society* 71 (2): 257 – 300.

Andrews, P. and Cook, J. 1985 Natural modifications to bones in a temperate setting. *Man* 20: 675 – 691.

Anton, D. 1984 Aspects of geomorphological evolution; paleosols and dunes in Saudi Arabia. In *Quaternary Period in Saudi Arabia. Sedimentological, Hydrogeological, Hydrochemical, Geomorphological, Geochronological and Climatological Investigations in Western Saudi Arabia.* Jado, A.R. and Zötl, J.G. (eds.) Wien & New York: Springer-Verlag pp 275 – 296.

Arribas, A. and Palmqvist, P. 1999 On the ecological connection between sabre-tooths and hominids: faunal dispersal events in the Lower Pleistocene and a review of the evidence for the first human arrival in Europe. *Journal of Archaeological Science* 26: 571 – 585.

Arthur, M.A., von Rad, U., Cornford, C., McCoy, F.W. and Sarnthein, M. 1979 Evolution and sedimentary history of the Cape Bojador continental margin, Northwestern Africa. In *Initial Reports of the Deep Sea Drilling Project, v. 47, Part 1.* von Rad, U. and Ryan, W.B.F. (eds.) Washington: U.S. Government Printing Office pp 773 – 816.

Asfaw, B., Beyene, Y., Semaw, S., White, T. and WoldeGabriel, G. 1991 Fejej: a new paleoanthropological research area in Ethiopia. *Journal of Human Evolution* 21: 137 – 143.

Ashfaque, S.M. and Saleem-ul-Haq 1991 Palaeolithic survey in Pabbi Hills field season: spring 1990. *Journal of Central Asia* 14 (1): 43 – 79.

Axelrod, D.I. 1944 The Sonoma flora (California). *Carnegie Institute of Washington Publication* 533: 167 – 206.

Azzaroli, A. and Napoleone, G. 1982 Magnetostratigraphic investigation of the Upper Sivaliks near Pinjor, India. *Rivista Italica Paleontologica* 87 (4): 739 – 762.

Badam, G.L. 1984 Pleistocene faunal succession of India. In *The Evolution of the East Asian Environment. Palaeobotany, Palaeozoology and Palaeoanthropology.* Volume II. Whyte, R.O. (ed.) Hong Kong: Centre of Asian Studies University of Hong Kong pp 746 – 775.

Ballog, R.A. and Malloy, R.E. 1981 Neogene palynology from the southern California continental borderland, Site 467, Deep Sea Drilling Project Leg 63. In *Initial Reports of the Deep Sea Drilling Project.* Volume 28. Yeats, R.S. and Haq, B.U., et al. (eds.) Washington, D.C.: U.S. Government Printing Office pp 565 – 576.

Bar-Yosef, O. 1994 The Lower Palaeolithic of the Near East. *Journal of World Prehistory* 8 (3): 211 – 265.

Bar-Yosef, O. 1995 The role of climate in the interpretation of human movements and cultural transformations in Western Asia. In *Paleoclimate and Evolution with Emphasis on Human Origins.* Vrba, E.S., Denton, G.H., Burckle, L.H. and Partridge, T.C. (eds.) New Haven & London: Yale University Press pp 507 – 523.

Bar-Yosef, O. and Belfer-Cohen, A. 2001 From Africa to Eurasia – early dispersals. *Quaternary International* 75: 19 – 28.

Barron, J.A. 1995 High resolution diatom paleoclimatology of the middle part of the Pliocene of the northwest Pacific. In *Proceedings of the Ocean Drilling Program, Scientific Results, 145.* Rea, D.K., Basov, I.A., Scholl, D.W. and Allan, J.F. (eds.) College Station, TX pp 43 – 53.

Barron, J.A. 1996a Diatom constraints on sea surface temperatures and sea ice distribution during the middle part of the Pliocene. *U.S. Geological Survey Open-File Report* 96 – 713: 1 – 45.

Barron, J.A. 1996b Diatom constraints on the position of the Antarctic Polar Front in the middle part of the Pliocene. *Marine Micropaleontology* 27: 195 – 213.

Barron, J.A., Cronin, T.M., Dowsett, H.J., Fleming, R.F., Holtz Jr, T.R., Ishman, S.E., Poore, R.Z., Thompson, R.S. and Willard, D.A. 1994 *Prism 8° x 10° Northern Hemisphere Paleoclimate Reconstruction: Digital Data.* http://geochange.er.usgs.gov/pub/ PRISM/OFR_94281/Derived/ report.html.

Barron, J.A, Cronin, C., Dowsett, H., Fleming, F., Holtz Jr, T., Ishman, S., Poore, R., Thompson, R. and Willard, D. 1995 Middle Pliocene paleoenvironments of the Northern Hemisphere. In *Paleoclimate and Evolution, with Emphasis on Human Origins.* Vrba, E.S., Denton, G.H., Partridge, T.C. and Burckle, L.H. (eds.) New Haven & London: Yale University Press pp 197 – 212.

Barry, J.C. 1995 Faunal turnover and diversity in the terrestrial Neogene of Pakistan. In *Paleoclimate and Evolution, with Emphasis on Human Origins.* Vrba, E.S., Denton, G.H., Partridge, T.C. and Burckle, L.H. (eds.) New Haven & London: Yale University Press pp 115 – 134.

Barry, J.C. and Flynn, L.J. 1990 Key biostratigraphic events in the Siwalik sequence. In *European Neogene Mammal Chronology.* Lindsay, E.H., Fahlbusch, V. and Mein, P. (eds.) New York: Plenum Press pp 557 – 571.

Bartstra, G.-J. 1982 *Homo erectus erectus*: the search for his artefacts. *Current Anthropology* 23 (3): 318 – 320.

Bartstra, G.-J. 1983a Comments and reply on 'The fauna from Trinil, type locality of *Homo erectus*: a reinterpretation'. Comment I: the vertebrate-bearing deposits of Kedungbrubus and Trinil, Java, Indonesia. *Geologie en Mijnbouw* 62 (2): 329 – 336.

Bartstra, G.-J. 1983b Some remarks upon fossil man from Java: his age, and his tools. *Bijdragen tot de Taal-Land-en Volkenkunde* 139: 421 – 434.

Baxter, M.J. 1994 *Exploratory Multivariate Analysis in Archaeology.* Edinburgh: Edinburgh University Press.

Beden, M. 1983 Family Elephantidae. In *Koobi Fora Research Project. Volume 2. The Fossil Ungulates: Proboscidea, Perissodactyla and Suidae.* Harris, J.M. (ed.) Oxford: Clarendon Press pp 40 – 129.

Behrensmeyer, A.K. 1975 Taphonomy and paleoecology in the hominid fossil record. *Yearbook of Physical Anthropology* 19: 36 – 50.

Behrensmeyer, A.K. 1978 The habitat of Plio-Pleistocene hominids in East Africa: taphonomic and microstratigraphic evidence. In *Early Hominids of Africa.* Jolly, C.J. (ed.) London: Duckworth. pp 165 – 189.

Behrensmeyer, A.K. 1985 Taphonomy and the palaeoecologic reconstruction of hominid habitats in the Koobi Fora Formation. In *L'Environment des hominidés au Plio-Pléistocène*. Coppens, Y. (ed.) Paris: Masson pp 309 – 324.

Behrensmeyer, A.K. and Cooke, H.B.S. 1985 Paleoenvironments, stratigraphy, and taphonomy in the African Pliocene and Early Pleistocene. In *Ancestors: The Hard Evidence*. E. Delson. (ed.) New York: Alan Liss pp 60 – 62.

Behrensmeyer, A.K. and Dechant Boaz, D.E. 1980 The recent bones of Amboseli National Park, Kenya, in relation to East African paleoecology. In *Fossils in the Making. Vertebrate Taphonomy and Paleoecology.* Behrensmeyer, A.K. and Hill, A.P. (eds.) Chicago: University of Chicago Press pp 72 – 92.

Behrensmeyer, A.K. and Kidwell, S.M. 1985 Taphonomy's contributions to paleobiology. *Paleobiology* 11 (1): 105 – 119.

Behrensmeyer, A.K., Western, D. and Dechant Boaz, D.E. 1979 New perspectives in vertebrate paleoecology from a recent bone assemblage. *Paleobiology* 5 (1): 12 – 21.

Bernet Kempers, A.J. 1982 Looking backward: pioneer prehistorians in pre-war Indonesia. *Modern Quaternary Research in South East Asia* 7: 1 – 23.

Bernor, R.L. 1983 Geochronology and zoogeographic relationship of Miocene Hominoidea. In *New Interpretations of Ape and Human Ancestry.* Ciochon, R.L. and Corruccini, R.S. (eds.) New York London: Plenum Press pp 21 – 64.

Bertolani Marchetti, D. 1975 III – preliminary palynological data on the proposed Plio-Pleistocene boundary type-section of Le Castella (1). *Ateneo Parmense, acta nat.* 11: 467 – 485.

Bertolani Marchetti, D., Accosi, C.A., Pelosio, G. and Raffi, S. 1979 Palynology and stratigraphy of the Plio-Pleistocene sequence of the Stirone River (northern Italy). *Pollen et Spores* 21: 149 – 167.

Beucher, F. 1975 *Étude palynologique de formations néogènes et quaternaires au Sahara nord-occidental.* Paris: C.N.R.S.

Bint, A.N. 1981 An Early Pliocene pollen assemblage from Lake Tay, south-western Australia, and its phytogeographic implications. *Australian Journal of Botany* 29: 277 – 291.

Bird, P. 1978 Finite element modelling of lithosphere deformation: the Zagros collision orogeny. *Tectonophysics* 50: 307 – 336.

Birks, H.J.B. and Birks, H.H. 1980 *Quaternary Palaeoecology.* London: Edward Arnold.

Birks, J.S. and Sinclair, C. 1982 The challenge of human resources development in Iraq. In *Iraq: The Contemporary State.* Niblock, T. (ed.) London & Canberra: Croom Helm pp 241 – 255.

Blisniuk, P.M., Hacker, B.R., Glodny, J., Ratschbachers, L., Siwen Bi, Zhenhan Wu, McWilliams, M.O. and Calvert, A. 2001 Normal faulting in Central Tibet since at least 13.5 Myr ago. *Nature* 412: 628 – 632.

Blood, P. 1994 *Pakistan: A Country Study.* http://lcweb2.loc.gov/cgi-bin/query/D?cstdy:1:./temp/ ~frd_tzuy::.

Boaz, N.T., Bernor, R.L., Brooks, A.S., Cooke, H.B.S., de Heinzelin, J., Dechamps, R., Delson, E., Gentry, A.W., Harris, J.W.K., Meylan, P., Pavakis, P.P., Sanders, W.J., Stewart, K.M., Verniers, J., Williamson, P.G. and Winkler, A.J. 1992 A new evaluation of the significance of the Late Neogene Lusso Beds, Upper Semliki Valley, Zaire. *Journal of Human Evolution* 22: 505 – 517.

Bonham-Carter, G.F., Agterberg, F.P. and Wright, D.F. 1989 Weights of evidence modelling: a new approach to mapping mineral potential. In *Statistical Applications in the Earth Sciences.* Volume 89 – 9. Agterberg, F.P. and Bonham-Carter, G.F. (eds.) Ottawa: Geological Survey of Canada pp 171 – 183.

Bonnefille, R. 1983 Evidence for a cooler and drier climate in the Ethiopian uplands towards 2.5 Myr ago. *Nature* 303: 487 – 491.

Bonnefille, R. 1985 Evolution of the continental vegetation: the palaeobotanical record from East Africa. *South African Journal of Science* 81: 267 – 270.

Bonnefille, R. 1995 A reassessment of the Plio-Pleistocene pollen record of East Africa. In *Paleoclimate and Evolution, with Emphasis on Human Origins.* Vrba, E.S., Denton, G.H., Partridge, T.C. and Burckle, L.H. (eds.) New Haven & London: Yale University Press pp 299 – 310.

Bonnefille, R., Vincens, A. and Buchet, G. 1987 Palynology, stratigraphy and paleoenvironment of a Pliocene hominid site (2.9 – 3.3 My) at Hadar, Ethiopia. *Palaeogeography, Palaeoclimatology, Palaeoecology* 60: 249 – 281.

Borisova, O.K. 1991 Neogene temperature fluctuations on the southeastern Russian Plain. Pliocene climates of the Northern Hemisphere. *U.S. Geological Survey Open-File Report* 91 – 447: 14 – 15.

Borisova, O.K. 1994 *Landscape and Climate of the South-Central and Southeastern Russian Plain in the Pliocene.* http://geochange.er.usgs.gov/pub/PRISM/OFR_94–23/15_Borisova.html.

Braches, F. and Shutler, R. 1983 Early vertebrates and the theory of the emergence of Java. *South East Asian Studies Newsletter* 13: 1 – 2.

Braches, F. and Shutler, R. 1984a Early vertebrates and the theory of the emergence of Java II. *South East Asian Studies Newsletter* 16: 1 – 2.

Braches, F. and Shutler, R. 1984b The Philippines and Pleistocene dispersal of mammals in island South East Asia. *Philippine Quarterly of Culture and Society* 12: 106 – 115.

Braches, F. and Shutler, R. 1984c Von Koenigswald's Cijulang fauna reconsidered. *Proceedings of the Sixth International Symposium on Asian Studies, 1984* 1: 831 – 848.

Brain, C.K. 1981 *The Hunters or the Hunted? An Introduction to African Cave Taphonomy.* Chicago: University of Chicago Press.

Bromage, T.G., Schrenk, F. and Juwayeyi, Y.M. 1995a Paleobiogeography of the Malawi Rift: age and vertebrate paleontology of the Chiwondo Beds, northern Malawi. *Journal of Human Evolution* 28: 37 – 57.

Bromage, T.G., Schrenk, F. and Zonneveld, F.W. 1995b Paleoanthropology of the Malawi Rift: an early hominid mandible from the Chiwondo Beds, northern Malawi. *Journal of Human Evolution* 28: 71 – 108.

Brouwers, E.M. 1994 Late Pliocene paleoecologic reconstructions based on ostracode assemblages from the Sagavanirktok and Gubik Formations, Alaskan North Slope. *Arctic* 47 (1): 16 – 33.

Brunet, M., Beauvilain, A., Coppens, Y., Heintz, E., Moutaye, A.H.E. and Pilbeam, D. 1995 The first australopithecine 2,500 kilometres west of the Rift Valley (Chad). *Nature* 378: 273 – 275.

Busche, D. 2001 Early Quaternary landslides of the Sahara and their significance for geomorphic and climatic history. *Journal of Arid Environments* 49: 429 – 448.

Butzer, K.W. and Hansen, C.L. with contributions by Leigh, E.G.J., Van Campo, M. and Gladfelter, B.G. 1968 *Desert and River in Nubia. Geomorphological and Prehistoric Environments at the Aswan Reservoir.* Madison: University of Wisconsin Press.

Caratini, C. and Tissot, C. 1988 Paleogeographical evolution of the Mahakam Delta in Kalimantan, Indonesia during the Quaternary and Late Pliocene. *Review of Palaeobotany and Palynology* 55: 217 – 228.

Carbonell, E., Bermúdez de Castro, J.M., Arsuaga, J.L., Diez, J.C., Rosas, A., Cuenca-Bescós, G., Sala, R., Mosquers, M. and Rodríguez, X.P. 1995 Lower Pleistocene hominids and artifacts from Atapuerca-TD6 (Spain). *Science* 269: 826 – 829.

Cerling, T.E. 1992 Development of grasslands and savannas in East Africa during the Neogene. *Palaeogeography, Palaeoclimatology, Palaeoecology (Global and Planetary Change Section)* 97: 241 – 247.

Cerling, T.E., Bowman, J.R. and O'Neil, J.R. 1988 An isotopic study of a fluvial-lacustrine sequence: the Plio-Pleistocene Koobi Fora sequence, East Africa. *Palaeogeography, Palaeoclimatology, Palaeoecology* 63: 335 – 356.

Cerling, T.E., Harris, J.M., MacFadden, B.J., Leakey, M.G., Quade, J., Eisenmann, V. and Ehleringer, J.R. 1997 Global vegetation change through the Miocene/Pliocene boundary. *Nature* 389: 153 – 158.

Cerling, T.E., Ehleringer, J.R. and Harris, J.M. 1998 Carbon dioxide starvation, the development of C_4 ecosystems, and mammalian evolution. *Philosophical Transactions of the Royal Society of London B* 353: 159 – 171.

Chandler, M. 1994 *GCM simulations of the Pliocene climate: feedbacks, ocean transports, and CO2.* http://geochange.er.usgs.gov/pub/PRISM/OFR_94–23/16_Chandler.html.

Chandler, M., Rind, D. and Thompson, R. 1994 Joint investigations of the Middle Pliocene climate II: GISS GCM Northern Hemisphere results. *Global and Planetary Change* 9: 197 – 219.

Chang, J. 1993 *Wild Swans.* London: Flamingo.

Chapin Metz, H. 1987 *Iran: A Country Study* http://lcweb2.loc.gov/cgi-bin/query/D?cstdy:15:./temp/~frd_OZy0::.

Chapman, R.W. 1978a General information on the Arabian Peninsula – geology. In *Quaternary Period in Saudi Arabia. Sedimentological, Hydrogeological, Hydrochemical, Geomorphological and Climatological Investigations in Central and Eastern Saudi Arabia.* Al-Sayari, S.S. and Zötl, J.G. (eds.) Wien & New York: Springer-Verlag pp 4 – 19.

Chapman, R.W. 1978b General information on the Arabian Peninsula – geomorphology. In *Quaternary Period in Saudi Arabia. Sedimentological, Hydrogeological, Hydrochemical, Geomorphological and Climatological Investigations in Central and*

Eastern Saudi Arabia. Al-Sayari, S.S. and Zötl, J.G. (eds.) Wien & New York: Springer-Verlag pp 19 – 30.

Chapman, R.W. 1978c Gulf region and its hinterland – geomorphology of the Eastern margin of the Shedgum Plateau. In *Quaternary Period in Saudi Arabia. Sedimentological, Hydrogeological, Hydrochemical, Geomorphological and Climatological Investigations in Central and Eastern Saudi Arabia.* Al-Sayari, S.S. and Zötl, J.G. (eds.) Wien & New York: Springer-Verlag pp 77 – 84.

Charlton, S.E.M. 1997 *Comparing Asian Politics. India, China and Japan.* Oxford: Westview Press.

Chi-yuen Wang, Yao-lin Shi and Wen-hu Zhou 1982 Dynamic uplift of the Himalaya. *Nature* 298: 553 – 556.

Choi, D.K. and Bong, P.Y. 1986 Neogene palynomorphs from lignite beds of Bugyeong and Yeonghae areas, Korea. *Journal of the Paleontological Society of Korea* 2: 1 – 17.

Church, T., Brandon, R.J. and Burgett, G.R. 2000 GIS applications in archaeology: method in search of theory. In *Practical Applications of GIS for Archaeologists. A Predictive Modeling Toolkit.* Wescott, K.L. and Brandon, R.J. (eds.) London: Taylor and Francis pp 135 – 155.

Ciochon, R.L., Larick, R., Zaim, Y., Sudijono and Heizler, M. 2001 Early Pleistocene dates and sedimentary cycles for new Bapang Formation hominins, Central Java, Indonesia. *American Journal of Physical Anthropology Annual Meeting Abstracts* Supplement 32: 50.

Clark, J.D. 1997 African Prehistory. In *History of Physical Anthropology*. Spencer, F. (ed.) New York & London: Garland Publishing, Inc. pp 32 – 41.

Clubb, O.E. 1964 *20th Century China.* New York & London: Columbia University Press.

Cohen, B. 1972 *Monsoon Asia – A Map Geography.* London: Heinemann Educational Books.

Coleman, M. and Hodges, K. 1995 Evidence for Tibetan plateau uplift before 14 Myr ago from a new minimum age for east-west extension. *Nature* 374 (6517): 49 – 52.

Coleman, R.G. 1974 Geological background of the Red Sea. *Initial Reports of the Deep Sea Drilling Project* 23: 813 – 819.

Combourieu-Nebout, N. 1995 Late Pliocene climate step and modifications of the vegetation in southern Italy from the high resolution pollen record of the Semaforo Section (2.4 – 2.1 Ma). *Comptes Rendus de L'academie des Sciences* 321 (8): 659 – 665.

Combourieu-Nebout, N. and Vergnaud Grazzini, C. 1991 Late Pliocene Northern Hemisphere glaciations: the continental and marine responses in the central Mediterranean. *Quaternary Science Reviews* 10: 319 – 334.

Conrad, G. 1969 L'évolution continentale post-hercynienne du Sahara Algérienne. *Ser. Geol. C.R.N.S.* 10: 1 – 527.

Cooke, H.B.S. 1978 Faunal evidence for the biotic setting of early African hominids. In *Early Hominids of Africa.* Jolly, C.J. (ed.) New York: St. Martin's pp 267 – 284.

Copeland, P. and Harrison, T.M. 1993 The Nyainqentanghla range, Southern Tibet: timing of the attainment of a large, high plateau and implications for development of the Asian monsoon. *Geological Society of America Abstracts Programme* 25: A–175.

Coppens, Y. and Howell, F.C. 1976 Mammalian faunas of the Omo Group: distributional and biostratigraphical aspects. In *Earliest Man and Environments in the Lake Rudolf Basin. Stratigraphy, Paleoecology, and Evolution.* Coppens, Y., Howell, F.C., Isaac, G.L. and Leakey, R.E.F. (eds.) Chicago & London: University of Chicago Press pp 177 – 192.

Cour, P. and Duzer, D. 1976 Persistance d'un climat hyperaride au Sahara central et méridional au cours de l'Holocène. *Revue de Gèographie Physique Gèologie Dynamique* 2 (18): 175 – 198.

Cravatte, J. and Suc, J.-P. 1981 Climatic evolution of North-Western Mediterranean area during Pliocene and early Pleistocene by pollen-analysis and forams of Drill Autan 1. Chronostratigraphic correlations. *Pollen et Spores* XXIII (2): 247 – 258.

Cronin, T.M. 1991a Neogene marine ostracoda from Tjörnes, Iceland. *Journal of Paleontology* 65 (5): 767 – 794.

Cronin, T.M. 1991b Pliocene shallow water paleoceanography of the North Atlantic Ocean based on marine ostracodes. *Quaternary Science Reviews* 10 (2 – 3): 175 – 188.

Cronin, T.M., Whatley, R.C., Wood, A., Tsukagoshi, A., Ikeya, N., Brouwers, E.M. and Briggs, W.M. 1993 Microfaunal evidence for elevated mid-Pliocene temperatures in the Arctic Ocean. *Paleoceanography* 8: 161 – 173.

Cronin, T.M., Kitamura, A., Ikeya, N., Watanabe, M. and Kamiya, T. 1994 Late Pliocene paleoceanography, Sea of Japan: The Yabuta Formation. *Palaeogeography, Palaeoclimatology, Palaeoecology* 108: 437 – 455.

Crossley, R., Watkins, C., Raven, M., Cripps, D., Carnell, A. and Williams, D. 1992 The sedimentary evolution of the Red Sea and Gulf of Aden. *Journal of Petroleum Geology* 15 (2): 157 – 172.

Crowley, T.J. 1996 Pliocene climates: the nature of the problem. *Marine Micropaleontology* 27: 3 – 12.

Dalla Bona, L. 2000 Protecting cultural resources through forest management planning in Ontario using archaeological predictive modeling. In *Practical Applications of GIS for Archaeologists. A Predictive Modeling Toolkit.* Wescott, K.L. and Brandon, R.J. (eds.) London: Taylor and Francis pp 73 – 99.

Day, M.H. 1976 Hominid postcranial remains from the East Rudolf Succession. In *Earliest Man and Environments in the Lake Rudolf Basin. Stratigraphy, Paleoecology, and Evolution.* Coppens, Y., Howell, F.C., Isaac, G.L. and Leakey, R.E.F. (eds.) Chicago & London: University of Chicago Press pp 507 – 521.

de Heinzelin, J., Clark, J.D., White, T., Hart, W., Renne, P., WoldeGabriel, G., Beyene, Y. and Vrba, E. 1999 Environment and behaviour of 2.5-million-year-old Bouri hominids. *Science* 284: 625 – 629.

de Vernal, A. and Mudie, P.J. 1989a Late Pliocene to Holocene palynostratigraphy at ODP Site 645, Baffin Bay. *Proceedings of the Ocean Drilling Program, Scientific Results* 105: 387 – 399.

de Vernal, A. and Mudie, P.J. 1989b Pliocene and Pleistocene palynostratigraphy at ODP Sites 646 and 647, eastern and southern Labrador Sea. *Proceedings of the Ocean Drilling Program, Scientific Results* 105: 401 – 422.

de Vos, J., Sartono, S., Hardja-Sasmiota, S. and Sondaar, P.Y. 1982 The fauna from Trinil, type locality of *Homo erectus*; a reinterpretation. *Geologie en Mijnbouw* 61: 207 – 211.

de Vos, J. and Sondaar, P. 1994 Dating hominid sites in Indonesia. *Science* 266: 1726 – 1727.

Delany, M.J. 1989 The zoogeography of the mammal fauna of Southern Arabia. *Mammal Review* 19 (4): 133 – 152.

DeMers, M.N. 1997 *Fundamentals of Geographic Information Systems.* New York: John Wiley & Sons, Inc.

Dennell, R.W. 1993 Evidence on human origins. A rediscovered source in the Upper Siwaliks of Northern Pakistan. *Interdisciplinary Science Reviews* 18 (4): 379 – 389.

Dennell, R.W. 2001 From Sangiran to Olduvai, 1937 – 1960: the quest for 'centres' of hominid origins in Asia and Africa. In *Studying Human Origins.* Corbey, R. and Roebroeks, W. (eds.) Amsterdam: Amsterdam University Press pp 45 – 66.

Dennell, R.W. in press *Plio-Pleistocene Investigations of the Pabbi Hills, Pakistan.* London: Kegan Paul.

Dennell, R.W., Rendell, H. and Hailwood, E. 1988 Early tool-making in Asia: two-million-year-old artefacts in Pakistan. *Antiquity* 62: 98 – 106.

Devyatkin, E.V., Dodonov, A.E., Gablina, S.S., Golovina, L.A., Kurenkova, V.G., Simakova, A.N., Trubikhin, V.M., Yasamanov, N.A., Khatib, K. and Nseir, H. 1996 Upper Pliocene-Lower Pleistocene marine deposits of Western Syria: stratigraphy and paleogeography. *Stratigraphy and Geological Correlation* 4 (1): 67 – 77.

Devyatkin, E.V., Dodonov, A.E., Sharkov, E.V., Zykin, V.S., Simakova, A.N., Khatib, K. and Nseir, H. 1997 The El-Ghab rift depression in Syria: its structure, stratigraphy, and history of development. *Stratigraphy and Geological Correlation* 5 (4): 362 – 374.

Dewey, J.F., Shackleton, R.M., Chang, C. and Sun, Y. 1988 The tectonic evolution of the Tibetan Plateau. *Philosophical Transactions of the Royal Society of London, A* 327: 379 – 413.

Ding, Z.L., Sun, J.M., Liu, T.S., Zhang, H.Z., Zhu, R.X., Yang, S.L. and Guo, B. 1988 Wind-blown origin of the Pliocene red clay formation in the Chinese Loess Plateau. *Earth and Planetary Science Letters* 161: 135 – 143.

Ding, Z.L., Xiong, S.F., Sun, J.M., Yang, S.L., Gu, Z.Y. and Liu, T.S. 1999 Pedostratigraphy and paleomagnetism of a ~7.0 Ma eolian loess-red clay sequence at Lingtai, Loess Plateau, north-central China and the implications for paleomonsoon evolution. *Palaeogeography, Palaeoclimatology, Palaeoecology* 152: 49 – 66.

Ding, Z.L., Rutter, N.W., Sun, J.M., Yang, S.L. and Liu, T.S. 2000 Re-arrangement of atmospheric circulation at about 2.6 Ma over northern China: evidence from grain size records of loess-paleosol and red clay sequences. *Quaternary Science Reviews* 19 (6): 547 – 558.

Ding, Z.L. and Yang, S.L. 2000 C_3/C_4 vegetation evolution over the last 7.0 Myr in the Chinese Loess Plateau: evidence from pedogenic carbonate $\delta^{13}C$. *Palaeogeography, Palaeoclimatology, Palaeoecology* 160: 291 – 299.

Ditchfield, D., Hicks, J., Plummer, T., Bishop, L.C. and Potts, R. 1999 Current research on the late Pliocene and Pleistocene deposits north of Homa Mountain, Southwestern Kenya. *Journal of Human Evolution* 36: 123 – 150.

Dodonov, A.E., Shackleton, N., Zhou, L.P., Lomov, S.P. and Finaev, A.F. 1999 Quaternary loess-paleosol stratigraphy of Central Asia: geochronology, correlation and evolution of paleoenvironments. *Stratigraphy and Geological Correlation* 7 (6): 581 – 593.

Dowsett, H.J. and Poore, R.Z. 1990 A new planktic foraminifer transfer function for estimating Pliocene through Holocene Sea Surface temperatures. *Marine Micropaleontology* 16 (1 – 2): 1 – 23.

Dowsett, H.J. and Poore, R.Z. 1991 Pliocene sea surface temperatures of the North Atlantic Ocean at 3.0 Ma. *Quaternary Science Reviews* 10 (2 – 3): 189 – 204.

Dowsett, H.J. and Loubere, P. 1992 High resolution Late Pliocene sea-surface temperature record from the Northeast Atlantic Ocean. *Marine Micropaleontology* 20: 91 – 105.

Dowsett, H.J. and Wiggs, L.B. 1992 Planktonic foraminiferal assemblage of the Yorktown Formation, USA. *Micropaleontology* 38: 75 – 86.

Dowsett, H.J., Thompson, R., Barron, J., Cronin, T., Fleming, F., Ishman, S., Poore, R., Willard, D. and Holtz Jr., T. 1994 Joint investigations of the Middle Pliocene climate I: PRISM paleoenvironmental reconstructions. *Global and Planetary Change* 9: 169 – 195.

Dowsett, H.J. and Ishman, S.E. 1995 Middle Pliocene planktic and benthic foraminifers from the sub-Arctic North Pacific: Sites 883 and 887. In *Proceedings of the Ocean Drilling Program, Scientific Results, 145*. Rea, D.K., Basov, I.A., Scholl, D.W. and Allan, J.F. (eds.) College Station, TX. pp 141 – 156.

Dowsett, H.J. and Robinson, M. 1998 Application of the modern analog technique (MAT) of sea surface temperature estimation to middle Pliocene North Pacific planktic foraminifer assemblages. *Paleontologia Electronica* 1 (1).

Dowsett, H.J., Barron, J.A., Poore, R.Z., Thompson, R.S., Cronin, T.M., Ishman, S.E. and Willard, D.A. 1999 *Middle Pliocene Paleoenvironmental Reconstruction: PRISM2.* http://pubs.usgs.gov/openfile/of99–535/.

Dubois, E. 1936 Nieuwe *Pithecanthropus* ontdekt? *Nieuwe Amsterdamsche Courant Algemeen Handelsblad* 35634 (10).

Duchaufour, P. 1977 *Pedology.* London: Allen and Unwin.

Duignan, P. and Gann, L.H. 1981 *The Middle East and North Africa. The Challenge to Western Security.* Stanford, Ca.: Hoover Institution Press.

Duncan, R.B. and Beckman, K.A. 2000 The application of GIS predictive site location models within Pennsylvania and West Virginia. In *Practical Applications of GIS for Archaeologists. A Predictive Modeling Toolkit.* Wescott, K.L. and Brandon, R.J. (eds.) London: Taylor and Francis pp 33 – 58.

Duowen, M. and Derbyshire, E. 1991 The depositional environment of the Late Pliocene 'red clay', Jing-Le Basin, Shanxi Province, China. *Sedimentary Geology* 70: 33 – 40.

Dupont, L.M. and Leroy, S. 1994 *Steps toward drier climatic conditions in North-Western Africa during the Upper Pliocene.* http://geochange.er.usgs.gov/pub/PRISM/OFR_94–23/11_Dupont.html.

Dupont, L.M. and Leroy, S.A.G. 1995 Steps towards drier climatic conditions in Northwestern Africa during the Upper Pliocene. In *Paleoclimate and Evolution, with Emphasis on Human Origins.* Vrba, E.S., Denton, G.H., Partridge, T.C. and Burckle, L.H. (eds.) New Haven & London: Yale University Press pp 289 – 298.

Ebert, J.I. 2000 The state of the art in 'inductive' predictive modeling: seven big mistakes (and lots of smaller ones). In *Practical Applications of GIS for Archaeologists. A Predictive Modeling Toolkit.* Wescott, K.L. and Brandon, R.J. (eds.) London: Taylor and Francis pp 129 – 134.

Edwards, M. 1992 Global gridded elevation and bathymetry. In *Global Ecosystems Database. Version 1.0 (on CD ROM). Documentation Manual, Disc A: National Geophysical Data Centre, Key to Geophysical Records Documentation No. 26. (Incorporated in: Global Change Database, Volume 1).* Kineman, J.J. and Ohrenschall, M.A. (eds.) Boulder, CO.: National Oceanic and Atmospheric Administration pp A14–11 – A14–14.

Eisenmann, V. 1983 Family Equidae. In *Koobi Fora Research Project. Volume 2. The Fossil Ungulates: Proboscidea, Perissodactyla and Suidae.* Harris, J.M. (ed.) Oxford: Clarendon Press pp 156 – 214.

El-Asmar, H.M. and Abdel-Fattah, Z.A. 2000 Lithostratigraphy and facies development of the Neogene-Quaternary succession in the Marsa Alam area, Red Sea coastal plain, Egypt. *Neues Jahrbuch für Geologie und Paläontologie-Abhandlungen* 217 (3): 397 – 431.

Environmental Systems Research Institute Inc 1992 – 1998 *ArcView GIS v3.1 Help Topics.*

Evernden, J.F. and James, G.T. 1964 Potassium-argon dates and the Tertiary floras of North America. *American Journal of Science* 262: 945 – 974.

Falcon, N.L. 1974 Southern Iran: Zagros Mountains. In *Mesozoic-Cenozoic Orogenic Belts. Data for Orogenic Studies.* Spencer, A.M. (ed.) Edinburgh: Scottish Academic Press pp 199 – 211.

Feibel, C.S., Brown, F.H. and McDougall, I. 1989 Stratigraphic context of fossil hominids from the Omo Group deposits: Northern Turkana Basin, Kenya and Ethiopia. *American Journal of Physical Anthropology* 78: 595 – 622.

Fernández-Jalvo, Y., Denys, C., Andrews, P., Williams, T., Dauphin, Y. and Humphrey, L. 1998 Taphonomy and palaeoecology of Olduvai Bed I (Pleistocene, Tanzania). *Journal of Human Evolution* 34 (2): 137 – 172.

Fielding, E., Isacks, B., Barazangi, M. and Duncan, C. 1994 How flat is Tibet? *Geology* 22 (2): 163 – 167.

Fieller, N.J. 2001 *Multivariate Data Analysis.* http://www.shef.ac.uk/nickfieller/shef-only/mvachapter2_dimensionalityreduction.pdf.

Fleming, R.F. 1992 Palynological data from Pliocene sediments, DSDP Leg 5 Site 32, Northeastern Pacific Ocean. *U.S. Geological Survey Open-File Report* 92 – 712: 24 pp.

Fleming, R.F. 1994 Cretaceous pollen in Pliocene rocks: implications for Pliocene climate in the southwestern United States. *Geology* 22: 787 – 790.

Fleming, R.F. and Barron, J.A. 1996 Evidence of Pliocene Notofagus in Antarctica from Pliocene marine sedimentary deposits (DSDP Site 274). *Marine Micropaleontology* 27: 227 – 236.

Foley, R.A. 1984 Early man and the red queen: tropical African community evolution and hominid adaptation. In *Hominid Evolution And Community Ecology: Prehistoric Human Adaptation In Biological Perspective.* Foley, R.A. (ed.) London: Academic Press pp 85 – 110.

Fort, M. 1996 Late Cenozoic environmental changes and uplift on the northern side of the central Himalaya: a reappraisal from field data. *Palaeogeography, Palaeoclimatology, Palaeoecology* 120 (1 – 2): 123 – 145.

Fort, M., Freytet, P. and Colchen, M. 1982 Structural and sedimentological evolution of the Thakkhola Mustang Graben (Nepal, Himalayas). *Zeitschrift für Geomorphologie N.E.* 42 (Suppl.): 75 – 98.

Fradkina, A.F. 1991 Pliocene climatic fluctuations in the far North-East of the USSR. Pliocene Climates of the Northern Hemisphere. *U.S. Geological Survey Open-File Report* 91 – 447: 22.

Frederick, W.H. and Worden, R.L. 1992 *Indonesia: A Country Study.* http://lcweb2.loc.gov/frd/cs/idtoc.html.

Freytet, P., Baltzer, F., Conchon, O., Plaziat, J.C. and Purser, B.H. 1994 Hydrologic significance of Quaternary continental carbonates from the Red Sea coast of the Egyptian Eastern Desert. *Bulletin de la Societé Géologique de France* 165 (6): 593 – 601.

Fuji, N. 1988 Palaeovegetation and palaeoclimate changes around Lake Biwa, Japan during the last ca. 3 million years. *Quaternary Science Reviews* 7: 21 – 28.

Gabunia, L. and Vekua, A. 1995 A Plio-Pleistocene hominid from Dmanisi, East Georgia, Caucasus. *Nature* 373: 509 – 512.

Gabunia, L., Vekua, A. and Lordkipanidze, D. 2000a The environmental contexts of early human occupation of Georgia (Transcaucasia). *Journal of Human Evolution* 38: 785 – 802.

Gabunia, L., Vekua, A., Lordkipanidze, D., Swisher, C.C., Ferrings, R., Justus, A., Nioradze, M., Tvalchrelidze, M., Antón, S., Bosinski, G., Jöris, O., de Lumley, M.A., Majsuradze, G. and Mouskhelishvili, A. 2000b Earliest Pleistocene hominid cranial remains from Dmanisi, Republic of Georgia: taxonomy, geological setting, and age. *Science* 288: 1019 – 1025.

Gamble, C. 1993 *Timewalkers.* Stroud: Alan Sutton.

Gaur, R. and Chopra, S.R.K. 1984 Taphonomy, fauna, environment and ecology of Upper Sivaliks (Plio-Pleistocene) near Chandigarh, India. *Nature* 308: 353 – 355.

Gazis, C.A., Lanphere, M., Taylor, H.P. and Gurbanov, A. 1995 Ar–40/Ar–39 and O–18/O–16 studies of the Chegem ash-flow caldera and the Eldjurta granite-cooling of two Late Pliocene igneous bodies in the Greater Caucasus Mountains, Russia. *Earth and Planetary Science Letters* 134 (3–4): 377 – 391.

Gibert, J., Gibert, L.L., Iglesias, A. and Maestro, E. 1998 Two 'Oldowan' assemblages in the Plio-Pleistocene deposits of the Orce region, southeast Spain. *Antiquity* 72: 17 – 25.

Gillings, M., Wise, A., (eds.) with contributions by Gillings, M., Halls, P., Lock, G., Miller, P., Phillips, G., Ryan, N., Wheatley, D. and Wise, A. 1999 *GIS Guide to Good Practice.* Oxford: Oxbow Books.

Gladenkov, Y.B., Barinov, K.B., Basilian, A.E. and Cronin, T.M. 1991 Stratigraphy and paleoceanography of Pliocene deposits of Karaginsky Island, eastern Kamchatka, U.S.S.R. *Quaternary Science Reviews* 10: 239 – 246.

Glennie, K.W. and Evamy, B.D. 1968 Dikaka: plants and plant-root structures associated with aeolian sand. *Palaeogeography, Palaeoclimatology, Palaeoecology* 4: 77 – 87.

Goldman, M. 1981 *China's Intellectuals. Advice and Dissent.* Cambridge, Mass.: Harvard University Press.

Golovina, L.A., Gablina, S.S. and Simakova, A.N. 1996 Paleogeographic interpretation of paleoflora from the Pliocene-Pleistocene sediments of Western Syria. *Stratigraphy and Geological Correlation* 4 (2): 202 – 205.

Graham, A. 1989 Late Tertiary paleoaltitudes and vegetational zonation in Mexico and Central America. *Acta Botanica Neerlandica* 38 (4): 417 – 424.

Graham, A. 1994 Neogene palynofloras and terrestrial paleoenvironments in northern Latin America. In *Pliocene Terrestrial Environments and Data/Model Comparisons.* U.S. Geological Survey Open-File Report 94 – 23. Thompson, R.S. (ed.) pp 23 – 30.

Green, S.W. 1990 Approaching archaeological space: an introduction to the volume. In *Interpreting Space: GIS and Archaeology.* Allen, K.M.S., Green, S.W. and Zubrow, E.B.W. (eds.) London: Taylor and Francis pp 3 – 8.

Greenacre, M.J. and Vrba, E.S. 1984 A correspondence analysis of biological census data. *Ecology* 65: 984 – 997.

Gregor, H.-J. 1990 Contributions to the Late Neogene and Early Quaternary history of the Mediterranean. *Review of Palaeobotany and Palynology* 62: 309 – 338.

Grichuk, V.P. 1991 Vegetation and climate of the middle Akchaghylian (Late Pliocene) on the Russian Plain. Pliocene Climates of the Northern Hemisphere. *U.S. Geological Survey Open-File Report* 91 – 447: 26 – 27.

Groot, J.J. 1991 Palynological evidence for late Miocene, Pliocene and early Pleistocene climate changes in the middle U.S. Atlantic coastal plain. *Quaternary Science Reviews* 10 (2 – 3): 147 – 162.

Grün, R. and Thorne, A. 1997 Dating the Ngandong humans. *Science* 276: 1575 – 1576.

Guo, S. 1981 On the elevation and climatic changes of the Qinghai-Xizang Plateau based on fossil angiosperms. In *Geological and Ecological Studies of the Qinghai-Xinzang (Tibet) Plateau.* Academia Sinica (ed.) Beijing & New York: Science Press, Gordon and Breach pp 201 – 206.

Guo Shengqiao and Shao Shixiong 1991 Quaternary lithofacies and paleogeography in China. In *The Quaternary of China.* Zhang Zonghu (ed.) China: China Ocean Press pp 122 – 158.

Hadiwisastra, S. 1984 The hominid-bearing Pucangan formation in Central and east Java. *Modern Quaternary Research in Southeast Asia* 8: 119 – 127.

Haeckel, E. 1868 *Natürliche Schöpfungsgeschichte.* Berlin: Reimer.

Haeckel, E. 1923 *The Evolution of Man.* London: Watts and Co.

Hageman, J.B. and Bennett, D.A. 2000 Construction of digital elevation models for archaeological applications. In *Practical Applications of GIS for Archaeologists. A Predictive Modeling Toolkit.* Wescott, K.L. and Brandon, R.J. (eds.) London: Taylor and Francis pp 113 – 127.

Han, J. 1995 *Climate Forcing: Observations from Chines Loess Plateau and a Perspective of the Influence of Oceanridge Processes.* London & Ontario: University of Western Ontario.

Han, J., Fyfe, W.S., Longstaffe, F.J., Palmer, H.C., Yan, F.H. and Mai, X.S. 1997 Pliocene-Pleistocene climatic change recorded in fluviolacustrine sediments in central China. *Palaeogeography, Palaeoclimatology, Palaeoecology* 135 (14): 27 – 39.

Hansen, D.T. 2000 *Describing GIS Applications: Spatial Statistics and Weights of Evidence Extension to ArcView in the Analysis of the Distribution of Archaeology Sites in the Landscape.* http://www.esri.com/library/userconf/proc00/professional/papers/PAP174/P174.htm.

Harris, J.M. 1978 Palaeontology. In *Koobi Fora Research Project Volume 1. The Fossil Hominids and an Introduction to Their Context.* Leakey, M.G. and Leakey, R.E. (eds.) Oxford: Clarendon Press pp 32 – 63.

Harris, J.M. 1983a Family Deinotheriidae. In *Koobi Fora Research Project. Volume 2. The Fossil Ungulates: Proboscidea, Perissodactyla and Suidae.* Harris, J.M. (ed.) Oxford: Clarendon Press pp 22 – 39.

Harris, J.M. 1983b Family Rhinocerotidae. In *Koobi Fora Research Project. Volume 2. The Fossil Ungulates: Proboscidea, Perissodactyla and Suidae.* Harris, J.M. (ed.) Oxford: Clarendon Press pp 130 – 155.

Harris, J.M. 1983c Family Suidae. In *Koobi Fora Research Project. Volume 2. The Fossil Ungulates: Proboscidea, Perissodactyla and Suidae.* Harris, J.M. (ed.) Oxford: Clarendon Press pp 215 – 302.

Harris, J.W.K., Isaac, G., Kaufulu, Z., Kroll, E., Schick, K., Toth, N. and Behrensmeyer, A. 1980 FxJj50: an Early Pleistocene site in Northern Kenya. *World Archaeology* 12: 109 – 136.

Harris, T. 2000 Session 2 discussion: moving GIS: exploring movement within prehistoric cultural landscapes using GIS. In *Beyond the Map: Archaeology and Spatial Technologies*. Lock, G. (ed.) Oxford: IOS Press pp 116 – 123.

Harrison, T.M., Copeland, P., Kidd, W.S.F. and An Yin 1992 Raising Tibet. *Science* 255: 1663 – 1670.

Hay, R.L. 1971 Geological background of Beds I and II stratigraphic summary. In *Olduvai Gorge Volume 3. Excavations in Beds I and II, 1960 – 1963*. Leakey, M.D. (ed.) Cambridge: Cambridge University Press pp 9 – 18.

Hay, R.L. 1994 Geology and dating of Beds III, IV and the Masek Beds. In *Olduvai Gorge Volume 5. Excavations in Beds III, IV and the Masek Beds, 1968 – 1971*. Leakey, M.D. (ed.) Cambridge: Cambridge University Press pp 8 – 14.

Hays, P.E., Pisias, N.G. and Roelofs, A.K. 1989 Paleoceanography of the eastern equatorial Pacific during the Pliocene: A high resolution study. *Paleoceanography* 4: 57 – 73.

Haywood, A.M., Valdes, P.J. and Sellwood, B.W. 2000 Global scale palaeoclimate reconstruction of the middle Pliocene climate using the UKMO GCM: initial results. *Global and Planetary Change* 25: 239 – 256.

Heitzman, J. and Worden, R.L. 1995 *India: A Country Study*. http://lcweb2.loc.gov/frd/cs/ intoc.html.

Heusser, L.E. 1981 Pollen analysis of selected samples from Deep Sea Drilling Project Leg 63. In *Initial Reports of the Deep Sea Drilling Project*. Yeats, R.S. and Haq, B.U., et al (eds.) Washington, D.C.: U.S. Government Printing Office pp 559 – 563.

Heusser, L.E. 1988 Pollen distribution in marine sediments on the continental margin off northern California. *Marine Geology* 80: 131 – 147.

Heywood, I., Cornelius, S. and Carver, S. 1998 *An Introduction to Geographical Information Systems*. Essex: Longman.

Hill, A., Ward, S., Deino, A., Curtis, G. and Drake, R. 1992 Earliest *Homo*. *Nature* 355: 719 – 722.

Hill, R.S. and MacPhail, M.K. 1985 A fossil flora from rafted Plio-Pleistocene mudstones at Regatta Point, Tasmania. *Australian Journal of Botany* 33: 497 – 517.

Hooghiemstra, H. 1994a Paleoclimatic conditions around 3 million year BP: pollen evidence from Colombia. In *Pliocene Terrestrial Environments and Data/Model Comparisons*. U.S. Geological Survey Open-File Report 94 – 23. Thompson, R.S. (ed.) pp 31 – 37.

Hooghiemstra, H. 1994b Pliocene-Quaternary floral migration, evolution of Northern Andean ecosystems and climatic change: implications from the closure of the Panamanian Isthmus. *Profil* 7: 413 – 425.

Hooghiemstra, H. and Sarmiento, G. 1991 Long continental pollen record from a tropical intermontane basin: Late Pliocene and Pleistocene history from a 540-meter core. *Episodes* 14: 107 – 115.

Hooijer, D.A. 1975 Quaternary mammals west and east of Wallace's line. *Netherlands Journal of Zoology* 25: 46 – 56.

Hooijer, D.A. and Kurtén, B. 1984 Trinil and Kedungbrubus: the *Pithecanthropus*-bearing fossil faunas of Java and their relative age. *Annales Zoologica Fennici* 21: 135 – 141.

Hoorn, C., Ohja, T. and Quade, J. 2000 Palynological evidence for vegetation development and climatic change in the Sub-Himalayan Zone (Neogene, Central Nepal). *Palaeogeography, Palaeoclimatology, Palaeoecology* 163: 133 – 161.

Horowitz, A. 1987 Subsurface palynostratigraphy and paleoclimates of the Quaternary Jordan Rift Valley fill, Israel. *Israel Journal of Earth Science* 36: 31 – 44.

Horowitz, A. 1988 Synthèse de la stratigraphie des paléoclimats Quaternaires du rift de la Mer Morte, Israël. *L'Anthropologie (Paris)* 92: 753 – 767.

Horowitz, A. 1989 Continuous pollen diagrams for the last 3.5 M.Y. from Israel: vegetation, climate and correlation with the oxygen isotope record. *Palaeogeography, Palaeoclimatology, Palaeoecology* 72: 63 – 78.

Horowitz, A. and Horowitz, M. 1985 Subsurface late Cenozoic palynostratigraphy of the Hula basin, Israel. *Pollen et Spores* 27 (3 – 4): 365 – 390.

Hötzl, H. 1978 As Sulb Plateau-Wadi Ad Dawasir and its hinterland, general geology. In *Quaternary Period in Saudi Arabia. Sedimentological, Hydrogeological, Hydrochemical, Geomorphological and Climatological Investigations in Central and Eastern Saudi Arabia*. Al-Sayari, S.S. and Zötl, J.G. (eds.) Wien & New York: Springer-Verlag pp 226 – 230.

Hötzl, H., Felber, H. and Zötl, J.G. 1978a As Sulb Plateau-Wadi Ar Rimah. In *Quaternary Period in*

Saudi Arabia. Sedimentological, Hydrogeological, Hydrochemical, Geomorphological and Climatological Investigations in Central and Eastern Saudi Arabia. Al-Sayari, S.S. and Zötl, J.G. (eds.) Wien & New York: Springer-Verlag pp 173 – 194.

Hötzl, H., Lippolt, H.J., Maurin, V., Moser, H. and Rauert, W. 1978b As Sulb Plateau-Wadi Ad Dawasir and its hinterland, Quaternary studies on the recharge area situated in crystalline rock regions. In *Quaternary Period in Saudi Arabia. Sedimentological, Hydrogeological, Hydrochemical, Geomorphological and Climatological Investigations in Central and Eastern Saudi Arabia.* Al-Sayari, S.S. and Zötl, J.G. (eds.) Wien & New York: Springer-Verlag pp 230 – 239.

Hötzl, H., Maurin, V. and Zötl, J.G. 1978c Gulf coastal region and its hinterland – geological history of the Al Hasa area since the Pliocene. In *Quaternary Period in Saudi Arabia. Sedimentological, Hydrogeological, Hydrochemical, Geomorphological and Climatological Investigations in Central and Eastern Saudi Arabia.* Al-Sayari, S.S. and Zötl, J.G. (eds.) Wien & New York: Springer-Verlag pp 58 – 77.

Hötzl, H. and Maurin, V. 1978 As Sulb Plateau – cuesta region of the Tuwayq Mountains, Wadi Birk. In *Quaternary Period in Saudi Arabia. Sedimentological, Hydrogeological, Hydrochemical, Geomorphological and Climatological Investigations in Central and Eastern Saudi Arabia.* Al-Sayari, S.S. and Zötl, J.G. (eds.) Wien & New York: Springer-Verlag pp 209 – 216.

Hötzl, H. and Zötl, J.G. 1978 Summary and general conclusions – climatic changes during the Quaternary period. In *Quaternary Period in Saudi Arabia. Sedimentological, Hydrogeological, Hydrochemical, Geomorphological and Climatological Investigations in Central and Eastern Saudi Arabia.* Al-Sayari, S.S. and Zötl, J.G. (eds.) Wien & New York: Springer-Verlag pp 264 – 311.

Hötzl, H. and Zötl, J.G. 1984 Geochronology and climate of the Quaternary – Middle and Early Pleistocene. In *Quaternary Period in Saudi Arabia. Sedimentological, Hydrogeological, Hydrochemical, Geomorphological, Geochronological and Climatological Investigations in Western Saudi Arabia.* Jado, A.R. and Zötl, J.G. (eds.) Wien & New York: Springer-Verlag pp 332 – 335.

Hou Yamei, Potts, R., Yuan Baoyin, Guo Zhengtang, Deino, A., Wang Wei, Clark, J., Xie Guangmao and Huang Weiwen 2000 Mid-Pleistocene Acheulean-like stone technology of the Bose Basin, South China. *Science* 287: 1622 – 1626.

Hourani, A. 1988 Societies and economies. Education. In *The Cambridge Encyclopedia of the Middle East and North Africa.* Mostyn, T. and Hourani, A. (eds.) Cambridge: Cambridge University Press pp 139 – 143.

Howell, F.C., Haesaerts, P. and de Heinzelin, J. 1987 Depositional environments, archaeological occurrences and hominids from Members E and F of the Shungura Formation (Omo basin, Ethiopia). *Journal of Human Evolution* 16: 665 – 700.

Hsü, J. 1976 On the paleobotanical evidence for continental drift and the Himalayan uplift. *Paleobotanist* 25: 131 – 142.

Hsü, J. 1983 Late Cretaceous and Cenozoic vegetation in China, emphasizing their connections with North America. *Annals of the Missouri Botanical Garden* 70: 490 – 508.

Hsü, K.J., Montadert, L., Bernoulli, D., Cita, M.B., Erickson, A., Garrison, R.E., Kidd, R.B., Mèlierés, F., Müller, C. and Wright, R. 1977 History of the Mediterranean salinity crisis. *Nature* 267: 399 – 403.

Huang Wanpo, Fang Quiren et al. 1991 Wushan hominid site. *Ocean* 1991: 192 – 211.

Huang Wanpo, Ciochon, R., Gu Yumin, Larick, R., Fang Qiren, Schwarcz, H., Yonge, C., de Vos, J. and Rink, W. 1995 Early *Homo* and associated artefacts from Asia. *Nature* 378: 275 – 278.

Huaya Lu, Xiaodong Liu, Fuqing Zhang, Zhisheng An and Dodson, J. 1999 Astronomical calibration of loess-paleosol deposits at Luochuan, central Chinese Loess Plateau. *Palaeogeography, Palaeoclimatology, Palaeoecology* 154: 237 – 246.

Huber, W. 1999 *Reproject Grids v4.0.* Merion Station, PA.: 1996 – 1999 Quantitative Decisions.

Huffman, O.F. 2001 Geologic context and age of the Perning/Mojokerto *Homo erectus*, East Java. *Journal of Human Evolution* 40: 353 – 362.

Hunt, C.O. 1989 The palynology and correlation of the Walton Crag (Red Crag Formation, Pliocene). *Journal of the Geological Society, London* 146: 743 – 745.

Huntley, B. 1993 The use of climate response surfaces to reconstruct paleoclimate from Quaternary pollen and plant microfossil data. *Philosophical Transactions of the Royal Society of London Series B Biological Sciences* 341: 215 – 224.

Hussain, S.T., van den Bergh, G.D., Steensma, K.J., de Visser, J.A., de Vos, J., Arif, M., van Dam, J., Sondaar, P.Y. and Malik, S.B. 1992 Biostratigraphy of the Plio-Pleistocene continental sediments (Upper Siwaliks) of the Mangla-Samwal Anticline, Azad Kashmir, Pakistan. *Proceedings Koninklijke*

Nederlandse Akademie van Wetenschappen 95 (1): 65 – 80.

Hyodo, M., Watanabe, N., Sunata, W., Susanto, E.E. and Wahyono, H. 1993 Magnetostratigraphy of hominid fossil bearing formations in Sangiran and Mojokerto, Java. *Anthropological Science* 101 (2): 157 – 186.

Igarashi, Y., Yoshida, M. and Tabata, H. 1988 History of vegetation and climate in the Kathmandu Valley. In *Palaeoclimatic and Palaeoenvironmental Changes in Asia During the Last 4 Million Years.* Jain, D.V.S., Agrawal, D.P., Sharma, P. and Gupta, S.K. (eds.) Indian National Science Academy pp 212 – 225.

Ikeya, N. and Cronin, T. 1993 Quantitative analysis of ostracoda and water masses around Japan: application to Pliocene and Pleistocene paleoceanography. *Micropaleontology* 39: 263 – 284.

Ilhan, E. 1974 Eastern Turkey. In *Mesozoic-Cenozoic Orogenic Belts. Data for Orogenic Studies.* Spencer, A.M. (ed.) Edinburgh: Scottish Academic Press pp 187 – 197.

Iossifova, Y.I. and Semenov, V.V. 1998 Climate-stratigraphy of the Pre-Tiglian-Bavelian analogues in Central Russia (the Don drainage basin). *Mededelingen Nederlands Instituut voor Toegepaste Geowetenschappen TNO* 60: 327 – 338.

Isaac, G.L. and Harris, J.W.K. 1978 Archaeology. In *Koobi Fora Research Project. Volume 1. The Fossil Hominids and an Introduction To Their Context 1968 – 1974.* Volume 1. Leakey, M.G. and Leakey, R.E. (eds.) Oxford: Clarendon Press pp 64 – 85.

Itihara, M., Watanabe, N., Kadar, D. and Kumai, H. 1994 Quaternary stratigraphy of the hominid fossil bearing formations in the Sangiran area, central Java. *Courier Forschungsinstitut Senckenberg* 171: 123 – 128.

Jablonski, N.G. 1993 Quaternary environments and the evolution of primates in East Asia, with notes on two new specimens of fossil Cercopithecidae from China. *Folia Primatologica* 60: 118 – 132.

Jablonski, N.G., Whitfort, M.J., Roberts-Smith, N. and Xu Qinqi, 2000 The influence of life history and diet on the distribution of catarrhine primates during the Pleistocene in eastern Asia. *Journal of Human Evolution* 39 (2): 131 – 157.

Jado, A.R. and Zötl, J.G. 1984 *Quaternary Period in Saudi Arabia. Sedimentological, Hydrogeological, Hydrochemical, Geomorphological, Geochronological and Climatological Investigations in Western Saudi Arabia.* Wien & New York: Springer-Verlag.

Jansson, C.R. 1966 Recent pollen spectra from the deciduous and coniferous-deciduous forests of Northeastern Minnesota: a study in pollen dispersal. *Ecology* 47: 804 – 825.

Jenkins, D.G. 1992a The paleogeography, evolution and extinction of Late Miocene-Pleistocene planktonic foraminifera from the southwest Pacific. In *Centenary of Japanese Micropaleontology.* Ishizaki, K. and Saito, T. (eds.) Tokyo: Terra Scientific Publishing Company pp 27 – 35.

Jenkins, D.G. 1992b Predicting extinctions of some extant planktic foraminifera. *Marine Micropaleontology* 19: 239 – 243.

Jian Wang, Yong Jin Wang, Ze Chun Liu, Jian Qing Li and Xi, P. 1999 Cenozoic environmental evolution of the Qaidam Basin and its implications for the uplift of the Tibetan Plateau and the drying of central Asia. *Palaeogeography, Palaeoclimatology, Palaeoecology* 152 (1 – 2): 37 – 47.

Johnson, D.H. 1978 Gulf coastal region and its hinterland – general geology. In *Quaternary Period in Saudi Arabia. Sedimentological, Hydrogeological, Hydrochemical, Geomorphological and Climatological Investigations in Central and Eastern Saudi Arabia.* Al-Sayari, S.S. and Zötl, J.G. (eds.) Wien & New York: Springer-Verlag pp 45 – 50.

Johnson, G.D., Rey, P., Ardrey, R.H., Visser, C.F., Opdyke, N.D. and Tahirkheli, R.A.K. 1981 Paleoenvironments of the Siwalik Group, Pakistan and India. In *Hominid Sites: Their Geologic Settings.* Jr., G.R. and Vondra, C.F. (eds.) Boulder, Colorado: Westview Press for the American Association for the Advancement of Science pp 197 – 254.

Johnson, G.D., Zeitler, P., Naeser, C.W., Johnson, N.M., Summers, D.M., Frost, C.D., Opdyke, N.D. and Tahirkheli, R.A.K. 1982a The occurrence and fission-track ages of Late Neogene and Quaternary volcanic sediments, Siwalik Group, Northern Pakistan. *Palaeogeography, Palaeoclimatology, Palaeoecology* 37: 63 – 93.

Johnson, N.M., Opdyke, N.D., Johnson, G.D., E.H., L. and Tahirkheli, R.A.K. 1982b Magnetic polarity stratigraphy and ages of Siwalik group rocks of the Potwar Plateau, Pakistan. *Palaeogeography, Palaeoclimatology, Palaeoecology* 37: 17 – 42.

Johnston, C.A. 1998 *Geographic Information Systems in Ecology.* Oxford: Blackwell Science.

Jule Xiao and Zhisheng An 1999 Three large shifts in East Asian monsoon circulation indicated by loess-paleosol sequences in China and late Cenozoic deposits in Japan. *Palaeogeography, Palaeoclimatology, Palaeoecology* 154: 179 – 189.

Junfeng Ji, Balsam, W. and Chen, J. 2001 Mineralogic and climatic interpretations of the Luochuan loess section (China) based on diffuse reflectance spectrophotometry. *Quaternary Research* 56: 23 – 30.

Kalb, J.E., Jolly, C.J., Mebrate, A., Tebedge, S., Smart, C., Oswald, E.B., Cramer, D., Whitehead, P., Wood, C.B., Conroy, G.C., Adefris, T., Sperling, L. and Kana, B. 1982 Fossil mammals and artefacts from the Middle Awash Valley, Ethiopia. *Nature* 298: 25 – 29.

Kamermans, H. 2000 Land evaluation as predictive modelling: a deductive approach. In *Beyond the Map: Archaeology and Spatial Technologies.* Lock, G. (ed.) Oxford: IOS Press pp 124 – 146.

Kappelman, J., Plummer, T., Bishop, L., Duncan, A. and Appleton, S. 1997 Bovids as indicators of Plio-Pleistocene palaeoenvironments in East Africa. *Journal of Human Evolution* 32: 229 – 256.

Kedves, M. 1984 Étude palynologique d'un lignite Tertiaire de Blao, Viet-Nam -1-. *Acta Biologica (Szeged)* 30: 91 – 105.

Keller, G. 1978 Late Neogene biostratigraphy and paleoceanography of DSDP Site 310 Central North Pacific and correlation with the Southwest Pacific. *Marine Micropaleontology* 3: 97 – 119.

Kemp, L.D., Bonham-Carter, G.F. and Raines, G.L. 1999 *Arc-WofE: ArcView extension for weights of evidence mapping.* http://ntserv.gis.nrcan.gc.ca/wofe.

Kemp, L.D., Bonham-Carter, G.F., Raines, G.L. and Looney, C.G. 2001 *Arc-SDM: ArcView extension for spatial data modelling using weights of evidence, logistic regression, fuzzy logic and neural network analysis.* http://ntserv.gis.nrcan.gc.ca/sdm.

Khan, A.M. 1974 Palynology of Neogene sediments from Papua (New Guinea) stratigraphic boundaries. *Pollen et Spores* 16: 265 – 284.

Kibunjia, M., Roche, H., Brown, F.H. and Leakey, R.E.F. 1992 Pliocene and Pleistocene archaeological sites west of Lake Turkana, Kenya. *Journal of Human Evolution* 23: 431 – 483.

Kidd, D.M. and Ritchie, M. 2000 Inferring the patterns and causes of geographic variation in *Ephippiger ephippiger* (Orthoptera, Tettigoniidae) using geographical information systems (GIS). *Biological Journal of the Linnean Society* 71: 269 – 295.

Klein, R.G. 1989 *The Human Career. Human Biological and Cultural Origins.* Chicago & London: University of Chicago Press.

Kohler, T.A. and Parker, S.C. 1986 Predictive models for archaeological resource location. In *Advances in Archaeological Method and Theory. Volume 9.* Schiffer, M.B. (ed.) London: Academic Press pp 397 – 452.

Koizumi, I. 1985 Late Neogene paleoceanography in the western North Pacific. In *Initial Reports of the Deep Sea Drilling Project, vol. 86.* Heath, G.R., Burckle, L.H. and et al. (eds.) Washington, D.C.: U.S. Govt. Printing Office pp 429 – 438.

Korotkevich, E.L. 1988 *Istoria formirovania hipparionovoj fauny Vostochnoj Evropy. (The History of Hipparion Fauna Formation in East Europe).* Kiev: Naukova Dumka.

Kostopoulos, D.S. and Koufos, G.D. 1994 The Plio-Pleistocene artiodactyls of Macedonia (Northern Greece) and their biostratigraphic significance; preliminary report. *Comptes Rendus de L'academie des Sciences. Series II.* 318: 1267 – 1272.

Kruuk, H. 1972 *The Spotted Hyena.* Chicago: University of Chicago Press.

Kuiper, J.A. and Wescott, K.L. 1999 *A GIS Approach for Predicting Prehistoric Site Locations.* http://www.esri.com/library/userconf/proc99/proceed/papers/pap378/p378.htm.

Kukla, G. and Cílek, V. 1996 Plio-Pleistocene megacycles: record of climate and tectonics. *Palaeogeography, Palaeoclimatology, Palaeoecology* 120: 171 – 194.

Kullmer, O., Sandrock, O., Abel, R., Schrenk, F. and T.G. Bromage, Y.M.J. 1999 The first *Paranthropus* from the Malawi Rift. *Journal of Human Evolution* 37: 121 – 127.

Kuna, M. 2000 Session 3 discussion: comments on archaeological prediction. In *Beyond the Map: Archaeology and Spatial Technologies.* Lock, G. (ed.) Oxford: IOS Press pp 180 – 186.

Kvamme, K.L. 1989 Geographic information systems in regional archaeological research and data management. In *Archaeological Method & Theory Volume 1.* Schiffer, M.B. (ed.) Tucson: University of Arizona Press pp 139 – 203.

Langbroek, M. and Roebroeks, W. 2000 Extraterrestrial evidence on the age of the hominids from Java. *Journal of Human Evolution* 38: 595 – 600.

Lankester, E.R. 1868 *'The History of Creation' or 'The Development of the Earth and its Inhabitants by the Action of Natural Causes'.* New York: Appleton.

Larick, R., Ciochon, R.L., Zaim, Y., Sudijono, Suminto, Rizal, Y. and Aziz, F. 2000 Lithostratigraphic context for Kln-1993.05-SNJ, a fossil colobine maxilla from Jokotingkir, Sangiran Dome. *International Journal of Primatology* 21 (4): 731 – 759.

Larick, R., Ciochon, R.L., Zaim, Y., Sudijono, Suminto, Rizal, Y., Aziz, F., Reagan, M. and Heizler, M. 2001 Early Pleistocene ^{40}Ar/ ^{39}Ar ages for Bapang Formation hominins, Central Jawa, Indonesia. *Proceedings of the National Academy of Sciences of the United States of America* 98 (9): 4866 – 4871.

Le Houérou, H.N. 1997 Climate, flora and fauna changes in the Sahara over the past 500 million years. *Journal of Arid Environments* 37 (4): 619 – 647.

Leakey, M. 1971a Appendix B. List of identified faunal remains from known stratigraphic horizons in Beds I and II. In *Olduvai Gorge. Volume 3. Excavations in Beds I and II, 1960 – 1963.* Leakey, M.D. (ed.) Cambridge: Cambridge University Press pp 290 – 294.

Leakey, M. 1971b Introduction. In *Olduvai Gorge. Volume 3. Excavations in Beds I and II, 1960 – 1963.* Leakey, M.D. (ed.) Cambridge: Cambridge University Press pp 1 – 8.

Leakey, M.D. 1971c Lower Bed I. Site DK and Site FLK NN: Level 4. In *Olduvai Gorge Volume 3. Excavations in Beds I and II, 1960 – 1963.* Leakey, M.D. (ed.) Cambridge: Cambridge University Press pp 21 – 39.

Leakey, M.D. 1994a Bed III. Site JK (Juma's Korongo). In *Olduvai Gorge. Volume 5. Excavations in Beds III, IV and the Masek Beds, 1968 – 1971.* Leakey, M.D. (ed.) Cambridge: Cambridge University Press pp 15 – 35.

Leakey, M.D. 1994b The fauna. In *Olduvai Gorge. Volume 5. Excavations in Beds III, IV and the Masek Beds, 1968 – 1971.* Leakey, M.D. (ed.) Cambridge: Cambridge University Press pp 130 – 145.

Lean, C.B., Hounslow, M.W., Vine, F.J., Harwood, G.M., Elvidge, L., Fisk, K., Kendall, A.C. and Montgomery, P. 1998 Magnetostratigraphy and sedimentary evolution of the Late Miocene to Early Pleistocene sediments, Quseir region, Egyptian Red Sea. *Geophysical Journal International* 133: 435 – 450.

Leopold, E.B. and Wright, V.C. 1985 Pollen profiles of the Plio-Pleistocene transition in the Snake River Plain, Idaho. Late Cenozoic history of the Pacific Northwest. *American Association for the Advancement of Science, Pacific Division.* : 323 – 348.

Leroy, S., Ambert, P. and Suc, J.-P. 1994 Pollen record of the Saint-Macaire Maar (Herault, Southern France) – a Lower Pleistocene glacial phase in the Languedoc coastal-plain. *Review of Palaeobotany and Palynology* 80 (1 – 2): 149 – 157.

Leroy, S. and Dupont, L. 1994 Development of vegetation and continental aridity in northwestern Africa during the Late Pliocene: the pollen record of ODP Site 658. *Palaeogeography, Palaeoclimatology, Palaeoecology* 109: 295 – 316.

Levin, N. and Horowitz, A. 1987 Palynostratigraphy of the Early Pleistocene QI palynozone in the Jordan-Dead Sea Rift, Israel. *Israel Journal of Earth Sciences* 36 (1 – 2): 45 – 58.

Lewis, M.E. 1997 Carnivoran paleoguilds of Africa: implications for hominid food procurement strategies. *Journal of Human Evolution* 32: 257 – 288.

Li Xiwen, Sun Xiangjun and Walker, D. 1984 The potential for palaeobotany in the explanation of China's plant geography. In *The Evolution of the East Asian Environment. Palaeobotany, Palaeozoology and Palaeoanthropology.* Whyte, R.O. (ed.) Hong Kong: Centre for Asian Studies University of Hong Kong pp 433 – 447.

Lindsay, E. 1997 Eurasian mammal biochronology: an overview. *Palaeogeography, Palaeoclimatology, Palaeoecology* 133 (3 – 4): 117 – 128.

Lindsay, E.H., Opdyke, N.D. and Fejfar, O. 1997 Correlation of selected late Cenozoic European mammal faunas with the magnetic polarity time scale. *Palaeogeography, Palaeoclimatology, Palaeoecology* 133 (3 – 4): 205 – 226.

Litwin, R.J. and Andrle, V.A.S. 1992 Modern palynomorph and weather census data from the U.S. Atlantic Coast (Continental Margin Program samples and selected NOAA weather stations). *U.S. Geological Survey Open-File Report* 92 – 263: 31 pp.

Liu, T.S. 1985 *Loess and the Environment.* Beijing: China Ocean Press.

Liu Tungsheng (ed.) 1991 *Quaternary Geology and Environment in China.* Beijing: Science Press.

Lock, G. (ed.) 2000 In *Beyond the Map: Archaeology and Spatial Technologies.* Oxford: IOS Press

Lock, G. 2000 Session 1 discussion: a particular view. In *Beyond the Map: Archaeology and Spatial Technologies.* Lock, G. (ed.) Oxford: IOS Press pp 60 – 64.

Lock, G. and Harris, T. 2000 Introduction: Return to Ravello. In *Beyond the Map: Archaeology and Spatial Technologies.* Lock, G. (ed.) Oxford: IOS Press pp xiii – xxv.

Luck, S. (ed.) 1999 *Philip's World Atlas and Encyclopedia.* London: George Philip Limited.

Lyman, R.L. 1994 *Vertebrate Taphonomy.* Cambridge: Cambridge University Press.

Mackensen, A. 1992 Neogene benthic foraminifers from the Southern Indian Ocean (Kerguelen Plateau): biostratigraphy and paleoecology. In *Proceedings Ocean Drilling Program, Scientific Results, vol. 120.* Wise, S.W.J., Schlich, R. and et al. (eds.) College Station, TX: Ocean Drilling Program pp 649 – 673.

Maglio, V.J. 1975 Pleistocene faunal evolution in Africa and Eurasia. In *After the Australopithecines.* Butzer, K.W. and Isaac, G.L. (eds.) The Hague: Mouton Publishers pp 419 – 476.

Maling, D.H. 1992 *Co-ordinate Systems and Map Projections.* Oxford: Pergamon Press.

Mamedov, A.V. 1991 The paleogeography of the Trans-Caucasus region during the Pliocene climatic optimum. Pliocene Climates of the Northern Hemisphere. *U.S. Geological Survey Open-File Report* 91 – 447: 28 – 31.

Marble, D.F. 1990 Geographic information systems: an overview. In *Introductory Readings in Geographic Information Systems.* Peuquet, D.J. and Marble, D.F. (eds.) London: Taylor and Francis pp 8 – 17.

Martínez Navarro, B. and Palmqvist, P. 1995 Presence of the African machairodont *Megantereon whitei* (Broom, 1937) (Felidae, Carnivora, Mammalia) in the Lower Pleistocene site of Venta Micena (Orce, Granada, Spain), with some considerations on the origin, evolution and dispersal of the genus. *Journal of Archaeological Science* 22 (4): 569 – 582.

Martínez Navarro, B. and Palmqvist, P. 1996 Presence of the African saber-toothed felid *Megantereon whitei* (Broom, 1937) (Mammalia, Carnivora, Machairodontinae) in Apollonia-1 (Mygdonia Basin, Macedonia, Greece). *Journal of Archaeological Science* 23 (6): 869 – 872.

Martínez Navarro, B., Turq, A., Ballester, J.A. and Oms, O. 1997 Fuente Nueva-3 (Orce, Granada, Spain) and the first human occupation of Europe. *Journal of Human Evolution* 33: 611 – 620.

Mathur, Y.K. 1984 Cenozoic palynofossils, vegetation, ecology and climate of the north and northwestern subhimalayan region, India. In *The Evolution of the East Asian Environment. Palaeobotany, Palaeozoology and Palaeoanthropology.* Whyte, R.O. (ed.) Hong Kong: Centre of Asian Studies University of Hong Kong pp 504 – 551.

Matsu'ura, S. 1982 A chronological framing for the Sangiran hominids: fundamental study by the Fluorine Dating method. *Bulletin of the National Science Museum Tokyo* Ser. D, 8: 1 – 53.

Matthews, E. 1985 Prescription of land-surface boundary conditions in GISS GCM II: a simple method based on high-resolution vegetation data bases. *NASA Rep.* TM 86096: 20 pp.

Matthews, J.V.J. 1987 Plant macrofossils from the Neogene Beaufort Formation on Banks and Meighen Islands, District of Franklin. *Current Research, Part A, Geological Survey of Canada* Paper 87–1A: 73 – 87.

Matthews, J.V.J. 1990 New data on Pliocene floras/faunas from the Canadian Arctic and Greenland. Pliocene climates: scenario for global warming. *U.S. Geological Survey Open-File Report* 90 – 64: 29 – 33.

Matthews, J.V.J. and Ovenden, L.E. 1990 Late Tertiary plant macrofossils from localities in Arctic/Subarctic North America: a review of the data. *Arctic* 43: 364 – 392.

McClure, H.A. 1978 As Sulb Plateau – Ar Rub' Al Khali. In *Quaternary Period in Saudi Arabia. Sedimentological, Hydrogeological, Hydrochemical, Geomorphological and Climatological Investigations in Central and Eastern Saudi Arabia.* Al-Sayari, S.S. and Zötl, J.G. (eds.) Wien & New York: Springer-Verlag pp 252 – 263.

Meisner, M. 1986 *Mao's China and After.* New York: The Free Press.

Michel, P. 1973 Les bassins des fleuves Sénégal et Gambie. Étude géomorphologique. *Mémoires Office de la Recherche Scientifique et Technique d'Outre-Mer* 63: 1 – 752.

Mildenhall, D.C. and Harris, W.F. 1970 A cool climate pollen assemblage from the type Waipipian (Middle Pliocene) of New Zealand. *New Zealand Journal of Geology and Geophysics* 13: 586 – 591.

Mildenhall, D.C. and Pocknall, D.T. 1983 Palaeobotanical evidence for changes in Miocene and Pliocene climates in New Zealand. In *Late Cainozoic Palaeoclimates of the Southern Hemisphere.* Vogel, J.C. (ed.) Rotterdam: A.A. Balkema pp 159 – 171.

Mildenhall, D.C. and Suggate, R.P. 1981 Palynology and age of the Tadmor Group (Late Miocene-Pliocene) and Porika Formation (Early Pleistocene), South Island, New Zealand. *New Zealand Journal of Geology and Geophysics* 24: 515 – 528.

Mitchell, J. and Westaway, R. 1999 Chronology of Neogene and Quaternary uplift and magmatism in the Caucasus: constraints from K-Ar dating of volcanism in Armenia. *Tectonophysics* 304 (3): 157 – 186.

Mohr, B.A.R. 1986 Die Mikroflora der Oberpliozänen tone Von Willershausen (Kreis Northeim, Niedersachsen). *Palaeontographica, Abt. B.* 198: 133 – 156.

Molnar, P. 1997 The rise of the Tibetan Plateau: from mantle dynamics to the Indian Monsoon. *Astronomy and Geophysics* 38 (3): 10 – 15.

Molnar, P. and England, P. 1990 Late Cenozoic uplift of mountain ranges and global climatic change: chicken or egg? *Nature* 346: 29 – 34.

Molnar, P., England, P. and Martinod, J. 1993 Mantle dynamics, uplift of the Tibetan Plateau, and the Indian monsoon. *Reviews of Geophysics* 31 (4): 357 – 396.

Monahan, C.M. 1996 New zooarchaeological data from Bed II, Olduvai Gorge, Tanzania: implications for hominid behaviour in the Early Pleistocene. *Journal of Human Evolution* 31 (2): 93 – 128.

Moore, P.D., J.A., W. and Collinson, M.E. 1991 *Pollen Analysis.* London: Blackwell Science.

Murakami, T. 1987 Orography and monsoons. In *Monsoons.* Fein, J.S. and Stephens, P.L. (eds.) New York: J. Wiley & Sons, Inc. pp 331 – 364.

Mworia, J., Dallmeijer, A. and Jacobs, B. 1988 Vegetation and modern pollen rain at Olorgesailie, Kenya. *Utafiti, Occasional Papers of the National Museums of Kenya.* 1: 1 – 22.

Nanda, A.C. 1994 Upper Siwalik mammalian faunas from Chandigarh and Jammu regions with comments on certain faunal discrepancies. In *Geology in South Asia – I.* Ahmed, R. and Sheikh, A.M. (eds.) Islamabad: Hydrocarbon Development Institute of Pakistan pp 39 – 45.

Nanda, A.C., Sati, D.C. and Mehra, G.S. 1991 Preliminary report on the stratigraphy and mammalian faunas of the Middle and the Upper Siwalik, west of River Yamuna, Paonta, Himachal Pradesh. *Journal of Himalayan Geology* 2 (2): 151 – 158.

Nelson, C.S., Mildenhall, D.C., Todd, A.J. and Pocknall, D.T. 1988 Subsurface stratigraphy, paleoenvironments, palynology, and depositional history of the Late Neogene Tauranga Group at Ohinewai, Lower Waikato Lowland, South Auckland. *New Zealand Journal of Geology and Geophysics* 31: 21 – 40.

Nelson, R.E. and Carter, L.D. 1985 Pollen analysis of a Late Pliocene and Early Pleistocene section from the Gubik Formation of arctic Alaska. *Quaternary Research* 24: 295 – 306.

Nemec, W. and Kazanci, N. 1999 Quaternary colluvium in west-central Anatolia: sedimentary facies and palaeoclimatic significance. *Sedimentology* 46: 139 – 170.

Ollivier, M.D. 1986 Contribution of [pollen] analysis to knowledge of [the] paleoenvironment in Spain. *Anthropologie Paris* 90: 589 – 591.

Opdyke, N.D., Lindsay, E., Johnson, G.D., Johnson, N., Tahirkheli, R.A.K. and Mirza, M.A. 1979 Magnetic polarity stratigraphy and vertebrate paleontology of the Upper Siwalik subgroup of Northern Pakistan. *Palaeogeography, Palaeoclimatology, Palaeoecology* 27: 1 – 34.

Orchiston, D.W. and Siesser, W.G. 1982 Chronostratigraphy of the Plio-Pleistocene fossil hominids of Java. *Modern Quaternary Research in S.E. Asia* 7: 131 – 149.

Orleans, L.A. 1960 *Professional Manpower and Education in Communist China.* Washington: National Science Foundation.

Ouda, K., Massoud, M. and Tammam, M. 2000 Stratigraphy of the Miocene sequence of the northern Red Sea. *Neues Jahrbuch für Geologie und Paläontologie – Abhandlungen* 215 (1): 125 – 176.

Oxford University Press 1993 *Oxford Atlas of the World.* New York: Oxford University Press.

Pagani, M., Freeman, K.H. and Arthur, M.A. 1999 Late Miocene atmospheric CO_2 concentrations and the expansion of C_4 grasses. *Science* 285: 876 – 879.

Palmqvist, P. 1997 A critical re-evaluation of the evidence for the presence of hominids in Lower Pleistocene times at Venta Micena, Southern Spain. *Journal of Human Evolution* 33 (1): 83 – 89.

Pan, Y. and Kidd, W.S.F. 1992 Nyainqentanglha shear zone – a late Miocene extensional detachment in the southern Tibetan Plateau. *Geology* 20 (9): 775 – 778.

Parés, J.M. and Pérez-González, A. 1995 Palaeomagnetic age for hominid fossils at Atapuerca archaeological site, Spain. *Science* 269: 830 – 832.

Parker Pearson, M. and Richards, C. 1994 *Architecture and Order.* London: Routledge.

Partridge, T.C., Wood, B.A. and deMenocal, P.B. 1995 Influence of global climatic change and regional uplift on large mammal evolution in east and southern Africa. In *Paleoclimate and Evolution with Emphasis on Human Origins.* Vrba, E.S., Denton, G.H., Burckle, L.H. and Partridge, T.C. (eds.) New Haven & London: Yale University Press pp 331 – 355.

Penrose, E. 1979 Industrial policy and performance in Iraq. In *The Integration of Modern Iraq.* Kelidar, A. (ed.) London: Croom Helm pp 150 – 170.

Pereira, J.M.C. and Itami, R.M. 1991 GIS-based habitat modeling using logistic multiple regression: a study of the Mt. Graham red squirrel. *Photogrammetric Engineering and Remote Sensing* 57 (11): 1475 – 1486.

Phillips, R.T. 1996 *China Since 1911.* London: MacMillan Press.

Pickford, M. 1990 Dynamics of old world biogeographic realms during the Neogene: implications for biostratigraphy. In *European Neogene Mammal Chronology.* Volume 180. Lindsay, E.H., Fahlbusch, V. and Mein, P. (eds.) New York: Plenum Press pp 413 – 442.

Planderová, E. 1974 The problem of the floristic boundary between Pliocene-Pleistocene in Western Carpathians mounts on the basis of palynological examination. *Bureau de Recherches Géologiques et Minières, Memoires* 78 (2): 547 – 551.

Plummer, T.W. and Bishop, L.C. 1994 Hominid paleoecology at Olduvai Gorge, Tanzania as indicated by antelope remains. *Journal of Human Evolution* 27: 47 – 75.

Plummer, T., Bishop, L.C, Ditchfield, P. and Hicks, J. 1999 Research on Late Pliocene Oldowan sites at Kanjera South, Kenya. *Journal of Human Evolution* 36: 151 – 170.

Poore, R.Z. 1999 Mid-Pliocene planktic foraminifers and environmental estimates DSDP Site 36. In *The Pliocene: Time of Change.* Wren, J.H., Suc, J.-P. and Leroy, S. (eds.) American Association of Stratigraphic Palynologists Special Publication XX, pp 198 – 208.

Poos, L.R., Wu, Y., Allen, J., Finan, T., Jewett, C. and Erdman, L. 1995 Digitizing history: GIS aids historical research. *GIS World* 8 (7): 48 – 51.

Pope, G.G. 1983 Evidence on the age of the Asian Hominidae. *Proceedings of the National Academy of Science, USA.* 80: 4988 – 4992.

Pope, G.G. 1997 Asian Paleoanthropology. In *History of Physical Anthropology.* Spencer, F. (ed.) New York & London: Garland Publishing, Inc. pp 127 – 133.

PRISM Project Members 1996 Pliocene planktic foraminifer census data from the North Atlantic region. *U.S. Geological Survey Open File Report* 96 – 669: 30 pp.

Qian Fang 1991 The Quaternary sediments and human fossils of Yuanmou Basin, Yunnan Province. In *Quaternary Geology and Environment in China.* Liu Tungsheng (ed.) Beijing: Science Press pp 351 – 357.

Quade, J., Cater, J.M.L., Ojha, T.P., Adam, J. and Harrison, T.M. 1995 Late Miocene environmental change in Nepal and the Northern Indian subcontinent: stable isotope evidence from paleosols. *Geological Society of American Bulletin* 107 (12): 1381 – 1397.

Quade, J., Cerling, T.E. and Bowman, J.R. 1989 Development of Asian monsoon revealed by marked ecological shift during the latest Miocene in northern Pakistan. *Nature* 342: 163 – 166.

Quade, J., Cerling, T.E., Bowman, J.R. and Jah, M.A. 1993 Palaeoecologic reconstruction of floodplain environments using palaeosols from Upper Siwalik Group sediments, Northern Pakistan. In *Himalaya to the Sea. Geology, Geomorphology and the Quaternary.* Shroder, J.F. (ed.) London & New York: Routledge pp 213 – 226.

Raines, G.L., Bonham-Carter, G.F. and Kemp, L. 2000 Predictive probabilistic modeling using ArcView GIS. *ArcUser* April – June 2000: 45 – 48.

Rand McNally 1991 *Rand McNally New International Atlas.* Chicago: Rand McNally.

Ravaut, P., Bayer, R., Hassani, R., Rousset, D. and Yahya'ey, A.A. 1997 Structure and evolution of the northern Oman margin: gravity and seismic constraints over the Zagros-Makran-Oman collision zone. *Tectonophysics* 279: 253 – 280.

Raynal, J.-P., Texier, J.-P., Geraads, D. and Sbihi-Alaoui, F.-Z. 1990 A new Plio-Pleistocene paleontological site in North Africa: Ahl Al Oughlam Quarry (former Deprez Quarry) at Casablanca, Morocco. *Comptes Rendus de L'académie des Sciences. Series II.* 310 (3): 315 – 320.

Reed, K. 1997 Early hominid evolution and ecological change through the African Plio-Pleistocene. *Journal of Human Evolution* 32: 289 – 322.

Reed, K.E. 1998 Using large mammal communities to examine ecological and taxonomic structure and predict vegetation in extant and extinct assemblages. *Paleobiology* 24 (3): 384 – 408.

Rendell, H.M., Dennell, R.W. and Halim, M. 1989 *Pleistocene and Palaeolithic Investigations in the*

Soan Valley, Northern Pakistan. Oxford: British Archaeological Reports.

Rendell, H.M., Hailwood, E.A. and Dennell, R.W. 1987 Magnetic polarity stratigraphy of Upper Siwalik Sub-Group, Soan Valley, Pakistan: implications for early human occupance of Asia. *Earth and Planetary Science Letters* 85: 488 – 496.

Robinson, H. 1976 *Monsoon Asia.* Plymouth: MacDonald and Evans.

Robinson, M.M. and Dowsett, H.J. 1996 Pliocene planktic foraminifer census data from DSDP Site 592, Southwestern Pacific Ocean. *U.S. Geological Survey Open-File Report* 96 – 544: 6 pp.

Roe, D.A. 1995 The Orce Basin (Andalucía, Spain) and the initial Palaeolithic of Europe. *Oxford Journal of Archaeology* 14 (1): 1 – 12.

Roebroeks, W. and Kolfschoten, V. 1994 The earliest occupation of Europe: a short chronology. *Antiquity* 68: 489 – 503.

Rognon, P. and Williams, M.A.J. 1977 Late Quaternary climatic change in Australia and North Africa: a preliminary interpretation. *Palaeogeography, Palaeoclimatology, Palaeoecology* 21: 285 – 327.

Rousseau, D.-D., Taoufiq, N.B., Petit, C., Farjanel, G., Méon, H. and Puisségur, J.-J. 1992 Continental late Pliocene paleoclimatic history recorded in the Bresse Basin (France). *Palaeogeography, Palaeoclimatology, Palaeoecology* 95: 253 – 261.

Sahni, A. and Mitra, H.C. 1980 Neogene palaeobiology of the Indian subcontinent with special reference to fossil vertebrates. *Palaeogeography, Palaeoclimatology, Palaeoecology* 31: 39 – 62.

Savage, S.H. 1990 GIS in archaeological research. In *Interpreting Space: GIS and Archaeology.* Allen, K.M.S., Green, S.W. and Zubrow, E.B.W. (eds.) London: Taylor and Francis pp 22 – 32.

Schrenk, F., Bromage, T.G., Betzier, C.G., Ring, U. and Juwayeyi, Y.M. 1993 Oldest *Homo* and Pliocene biogeography of the Malawi Rift. *Nature* 365: 833 – 836.

Schwarzbach, M. and Pflug, H.D. 1957 Das Klima des jüngeren Tertiärs in Island. *Neues Jahrbuch Geologie und Paläontologie* 104: 279 – 298.

Scott, L. 1984 Palynological evidence for Quaternary paleoenvironments in Southern Africa. In *Southern African Prehistory and Paleoenvironments.* Klein, R.G. (ed.) Rotterdam: A.A. Balkema pp 65 – 80.

Scott, L. and Partridge, T.C. 1994 Some manifestations of Pliocene warming in southern Africa. In *Pliocene Terrestrial Environments and Data/Model Comparisons.* U.S. Geological Survey Open-File Report 94 – 23. Thompson, R.S. (ed.) pp 54 – 55.

Searle, M. 1995 Plate tectonics: the rise and fall of Tibet. *Nature* 374 (6517): 17 – 18.

Searle, M.P. and Owen, L.A. 1999 The evolution of the Indus River in relation to topographic uplift, erosion, climate and geology of Western Tibet, the Trans-Himalayan and High Himalayan Ranges. In *The Indus River: Biodiversity, Resources, Humankind.* Meadows, A. and Meadows, P. (eds.) Pakistan: Oxford University Press .

Sémah, F. 1986 Le peuplement ancien de Java ébauche d'un cadre chronologique. (Early Javanese settlement: towards a chronological outline). *L'Anthropologie* 90 (3): 359 – 400.

Sémah, F., Saleki, H., Falguères, C., Féraud, G. and Djubiantono, T. 2000 Did early man reach Java during the Late Pliocene? *Journal of Archaeological Science* 27: 763 – 769.

Sept, J.M. 1986 Plant foods and early hominids at site FxJj 50, Koobi Fora, Kenya. *Journal of Human Evolution* 15: 751 – 770.

Shackleton, N.J., Backman, J., Zimmerman, H., Kent, D.V., Hall, M.A., Roberts, D.G., Schnitker, D., Baldauf, J.G., Desprairies, A., Homrighausen, R., Huddlestun, P., Keene, J.B., Kaltenback, A.J., Krumsiek, K.A.O., Morton, A.C., Murray, J.W. and Westberg-Smith, J. 1984 Oxygen isotope calibration of the onset of ice-rafting and history of glaciation in the North Atlantic region. *Nature* 307: 620 – 623.

Shackleton, R.M. and Chang, C. 1988 Cenozoic uplift and deformation of the Tibetan Plateau: the geomorphological evidence. *Philosophical Transactions of the Royal Society of London, A* 327: 365 – 377.

Shahar, J. 1994 The Syrian arc system: an overview. *Palaeogeography, Palaeoclimatology, Palaeoecology* 112: 125 – 142.

Shatilova, I.I. 1980 *Palynologic Study of Late Cainozoic and Modern Deposits in the Eastern Part of the Black Sea Area.* IV International Palynological Conference, Lucknow, India.

Shatilova, I.I. 1986 The palynological base of stratigraphical subdivision of Late Cainozoic deposits of the western Transcaucasus. *Review of Palaeobotany and Palynology* 48: 409 – 414.

Shatilova, I.I., Macharadze, N.V. and Davitashivilli, L.S. 1991 The Pliocene climate of western Georgia. Pliocene climates of the Northern Hemisphere. *U.S.*

Geological Survey Open-File Report 91 – 447: 35 – 37.

Shennan, S. 1997 *Quantifying Archaeology.* Edinburgh: Edinburgh University Press.

Shi, N., Cao, J.X. and Konigsson, L.K. 1993 Late Cenozoic vegetational history and the Pliocene-Pleistocene boundary in the Yushe Basin, SE Shanxi, China. *Grana* 32 (4 – 5): 260 – 271.

Shipman, P. and Harris, J.M. 1991 Habitat preference and paleoecology of *Australopithecus boisei* in Eastern Africa. In *Evolutionary History of the 'Robust' Australopithecines.* Grine, F. (ed.) New York: Aldine de Gruyter pp 343 – 381.

Shutler, R. and Braches, F. 1985 Problems in paradise: the Pleistocene of Java revisited. *Modern Quaternary Research in Southeast Asia* 9: 87 – 97.

Singh, G. 1988 History of aridland vegetation and climate: a global perspective. *Biological Review* 63: 159 – 195.

Sloan, L.C., Crowley, T.J. and Pollard, D. 1996 Modeling of middle Pliocene climate with the NCAR GENESIS general circulation model. *Marine Micropaleontology* 27: 51 – 61.

Smith, P.E.L. 1986 *Palaeolithic Archaeology in Iran.* Philadelphia: The University Museum.

Sondaar, P.Y. 1981 The Geochelone faunas of the Indonesian archipelago and their paleographical and biostratigraphical significance. *Modern Quaternary Research in S.E. Asia* 6: 111 – 120.

Sondaar, P.Y. 1984 Faunal evolution and the mammalian biostratigraphy of Java. *Courier Forschungsinstitut Senckenberg* 69: 219 – 235.

Sousa, A. 1982 The eradication of illiteracy in Iraq. In *Iraq: The Contemporary State.* Niblock, T. (ed.) London & Canberra: Croom Helm pp 100 – 108.

Spaak, P. 1983 Accuracy in correlation and ecological aspects of the planktonic foraminiferal zonation of the Mediterranean Pliocene. *Utrecht Micropaleontological Bulletins* 28: 159 pp.

Spassov, N. 1998 A new Late Villafranchian locality of vertebrate fauna – Slivnitsa (Bulgaria) and the carnivore dispersal events in Europe on the Plio-Pleistocene boundary. *Historia Naturalis Bulgarica* 9: 101 – 113.

Spencer, L.M. 1997 Dietary adaptations of Plio-Pleistocene Bovidae: implications for hominid habitat use. *Journal of Human Evolution* 32: 201 – 228.

SPSS for Windows 2000 *SPSS for Windows v. 10.0.7 Help Topics.*

Stančič, Z. and Veljianovski, T. 2000 Understanding Roman settlement patterns through multivariate statistics and predictive modelling. In *Beyond the Map: Archaeology and Spatial Technologies.* Lock, G. (ed.) Oxford: IOS Press pp 147 – 156.

Stern, N. 1993 The structure of the Lower Pleistocene archaeological record. *Current Anthropology* 34 (3): 201 – 225.

Stern, N. 1994 The implications of time-averaging for reconstructing the land-use patterns of early tool-using hominids. *Journal of Human Evolution* 27: 89 – 105.

Stine, R.S. and Decker, T.D. 1990 Archaeology, data integrations and GIS. In *Interpreting Space: GIS and Archaeology.* Allen, K.M.S., Green, S.W. and Zubrow, E.B.W. (eds.) London: Taylor and Francis pp 134 – 140.

Stoms, D.M., Davis, F.W. and Cogan, C.B. 1992 Sensitivity of wildlife habitat models to uncertainties in GIS data. *Photogrammetric Engineering and Remote Sensing* 58 (6): 843 – 850.

Straus, L.G. 2001 Africa and Iberia in the Pleistocene. *Quaternary International* 75: 91 – 102.

Straus, L.G. and Bar-Yosef, O. (eds) 2001 Out of Africa in the Pleistocene. *Quaternary International* 75

Street, A. and Gasse, F. 1981 Recent developments in research into the Quaternary climatic history of the Sahara. In *The Sahara: Ecological Change and Early Economic History.* Allan, J.A. (ed.) Cambridge: Mensa Press Ltd pp 7 – 28.

Stuchlik, L. and Shatilova, I.I. 1987 Palynological study of Neogene deposits of southern Poland and western Georgia. *Acta Palaeobotanica* 27 (2): 21 – 52.

Suc, J.-P. 1984 Origin and evolution of the Mediterranean vegetation and climate in Europe. *Nature* 307: 429 – 432.

Suc, J.P. and Zagwijn, W.H. 1983 Plio-Pleistocene correlations between the northwestern Mediterranean region and northwestern Europe according to recent biostratigraphic and paleoclimatic data. *Boreas* 12: 153 – 166.

Svetlitskaya, T.V. 1994 *Landscape and Climate of the Southwestern Russian Plain in the Pliocene.* http://geochange.er.usgs.gov/pub/PRISM/OFR_94–23/14_Svetlitskaya.html.

Swisher, C.C. 1997 A revised geochronology for the Plio-Pleistocene hominid bearing strata of Sangiran

Java, Indonesia. *Journal of Human Evolution* 32 (4): A23.

Swisher, C.C., Curtis, G.H., Jacob, T., Getty, A.G., Suprijo, A. and Widiasmoro. 1994 Age of the earliest known hominids in Java, Indonesia. *Science* 263: 1118 – 1121.

Swisher, C.C., Rink, W.J., Antón, S.C., Schwarcz, H.P., Curtis, G.H., Suprijo, A. and Widiasmoro 1996 Latest *Homo erectus* of Java: potential contemporaneity with *Homo sapiens* in Southeast Asia. *Science* 274: 1870 – 1874.

Tanai, T. and Huzioka, K. 1967 Climatic implications of Tertiary floras in Japan. In *Tertiary Correlations and Climatic Changes in the Pacific.* pp 89 – 94.

Tandon, S.K., Kumar, R., Koyama, M. and Niitsuma, N. 1984 Magnetic polarity stratigraphy of the Upper Siwalik subgroup, east of Chandigarh, Punjab sub-Himalaya, India. *Journal of the Geological Society of India* 25 (1): 45 – 55.

Tchernov, E. 1987 The age of the 'Ubeidiya Formation, an Early Pleistocene Hominid site in the Jordan Valley, Israel. *Israel Journal of Earth Sciences* 36: 3 – 30.

Tchernov, E. 1988 The paleobiogeographical history of the Southern Levant. In *The Zoogeography of Israel: the Distribution and Abundance at a Zoogeographical Crossroad.* Yom-Tov, Y. and Tchernov, E. (eds.) Dordrecht & Boston & Lancaster: Dr. W. Junk Publishers pp 159 – 250.

Tchernov, E. 1992 Eurasian-African biotic exchanges through the Levantine corridor during the Neogene and Quaternary. *Courier Forschungsinstitut Senckenberg* 153: 103 – 123.

Tchernov, E. 1994 New comments on the biostratigraphy of the Middle and Upper Pleistocene of the Southern Levant. In *Late Quaternary Chronology and Paleoclimates of the Eastern Mediterranean.* Bar-Yosef, O. and Kra, R.S. (eds.) Tucson, Ariz., USA & Cambridge, Mass., USA: RADIOCARBON; American School of Prehistoric Research pp 333 – 350.

Tchernov, E., Horwitz, L.K., Ronen, A. and Lister, A. 1994 The faunal remains from Evron Quarry in relation to other Lower Paleolithic hominid sites in the Southern Levant. *Quaternary Research* 42 (3): 328 – 339.

Teaford, M.F. and Walker, A.C. 1984 Quantitative differences in dental microwear between primate species with different diets and a comment on the presumed diet of *Sivapithecus. American Journal of Physical Anthropology* 64: 191 – 200.

The Evergreen Project and Collaborative, T.M.P.G. 1998 *Deserts of the World* http://www.mobot.org/pfg/diverse/biomes/desert/world.htm.

The State of Israel 1999 *Guide to the Mideast Peace Process.* http://www.mfa.gov.il/mfa/.

The Times 1990 *The Times Atlas of the World.* Edinburgh: Bartholomew and Times Books.

Thomas, H. 1984 Les bovidae (Artiodactyla: Mammalia) du Miocene du sous-continent Indien, de la peninsule Arabique et de l'Afrique: biostratigraphie, biogeographie et ecologie. *Palaeogeography, Palaeoclimatology, Palaeoecology* 45: 251 – 299.

Thomas, H., Geraads, D., Janjou, D., Vaslet, D., Memesh, A., Billiou, D., Bocherens, H., Dobigny, G., Eisenmann, V., Gayet, M., de Lapparent De Broin, F., Petter, G. and Halawani, M. 1998 First Pleistocene faunas from the Arabian Peninsula: An Nafud desert, Saudi Arabia. *Comptes Rendus de L'academie des Sciences. Sciences de la terre et des planètes* 326: 145 – 152.

Thompson, O.E. 1998 http://www.meto.umd.edu/~owen/CHPI/IMAGES/circs02.html.

Thompson, R.S. 1991 Pliocene environments and climates in the western United States. *Quaternary Science Reviews* 10 (2 – 3): 115 – 132.

Thompson, R.S. 1992 Palynological data from a 989 ft. (301 m) core of Pliocene and Pleistocene sediments from Bruneau, Idaho. *U.S. Geological Survey Open-File Report* 92 – 713: 28 pp.

Thompson, R.S. 1996 Pliocene and Early Pleistocene environments and climates of the western Snake River Plain, Idaho. *Marine Micropalaeontology* 27: 141 – 156.

Thompson, R.S. and Fleming, R.F. 1996 Middle Pliocene vegetation: reconstructions, paleoclimatic inferences, and boundary conditions for climate modeling. *Marine Micropaleontology* 27: 27 – 49.

Thompson, R.S., Oviatt, C.G., Roberts, A.P., Buchner, J., Kelsey, R., Bracht, C., Forester, R.M. and Bradbury, J.P. 1995 Stratigraphy, sedimentology, paleontology, and paleomagnetism of Pliocene-early Pleistocene lacustrine deposits in two cores from western Utah. *U.S. Geological Survey Open-File Report* 95 – 1: 94 pp.

Thunell, R.C. 1979a Climatic evolution of the Mediterranean Sea during the last 5.0 million years. *Sedimentary Geology* 23: 67 – 79.

Thunell, R.C. 1979b Pliocene-Pleistocene paleotemperature and paleosalinity history of the

Mediterranean Sea: Results from DSDP Sites 125 and 132. *Marine Micropaleontology* 4: 173 – 187.

Tiedemann, R. 1991 *Acht Millionen Jahre Klimageschichte von Nordwest Afrika und Paläo - Ozeanographie des angrenzenden Atlantiks: Hochauflösende Zeitreihen von ODP – Sites 658 – 661.* Kiel: University of Kiel.

Tiedemann, R., Sarnthein, M. and Stein, R. 1989 Climatic changes in the western Sahara: aeolo-marine sediment record of the last 8 million years (sites 657 – 661). *Proc. ODP, Sci. Results* 108: 241 – 277.

Tiedmann, R., Sarnthein, M. and Shackleton, N.J. 1994 Astronomic timescale for the Pliocene $\delta^{18}O$ and dust records of Ocean Drilling Program Site 659. *Paleoceanography* 9 (4): 619 – 638.

Tobias, P.V. 1991 *Olduvai Gorge. Volume 4. The Skulls, Endocasts and Teeth of Homo habilis.* Cambridge: Cambridge University Press.

Tomlinson, R.F. 1990 Current and potential uses of geographic information systems: the North American experience. In *Introductory Readings in Geographic Information Systems.* Peuquet, D.J. and Marble, D.F. (eds.) London: Taylor and Francis pp 142 – 158.

Tong Guobang and Shao Shixiong 1991 The evolution of Quaternary climate in China. In *The Quaternary of China.* Zhang Zonghu (ed.) China: China Ocean Press pp 42 – 76.

Traverse, A. 1982 Response of world vegetation to Neogene tectonic and climatic events. *Alcheringa* 6: 197 – 209.

Tsoukala, E. 1992 Quaternary faunas of Greece. *Courier Forschungsinstitut Senckenberg* 153: 79 – 92.

Tungsheng Liu, Zhongli Ding and Rutter, N. 1999 Comparison of Milankovitch periods between continental loess and deep sea records over the last 2.5 Ma. *Quaternary Science Reviews* 18: 1205 – 1212.

Turnbull, I.M., Lindqvist, J.K., Mildenhall, D.C., Hornibrook, N.d.B. and Beu, A.B. 1985 Stratigraphy and paleontology of Pliocene-Pleistocene sediments on Five Fingers Peninsula, Dusky Sound, Fiordland. *New Zealand Journal of Geology and Geophysics* 28: 217 – 231.

Turner, A. 1982 Hominids and fellow travellers. *South African Journal of Science* 78: 231 – 237.

Turner, A. 1992 Large carnivores and earliest European hominids: changing determinants of resource availability during the Lower and Middle Pleistocene. *Journal of Human Evolution* 22: 109 – 126.

Turner, A. 1999 Assessing earliest human settlement of Eurasia: Late Pliocene dispersions from Africa. *Antiquity* 73: 563 – 570.

Turner, A. in prep Vertebrate fossils. Carnivora. In *British Archaeological Mission to Pakistan.* Dennell, R.W. and et al. (eds.) pp 1 – 17.

Van Andel, T.H. 1981 Consider the incompleteness of the geological record. *Nature* 294: 397 – 398.

Van Bemmelen, R.W. 1949 *The Geology of Indonesia. Volume IA. General Geology of Indonesia and Adjacent Archipelagoes.* The Hague: Government Printing Office.

Van Campo, M. 1975 Pollen analysis in the Sahara. In *Problems in Prehistory: North Africa and the Levant.* Wendorf, F. and Marks, A.E. (eds.) Dallas: SMU Press pp 45 – 64.

Van Campo, M., Guinet, P. and Cohen, J. 1968 Fossil pollen from Late Tertiary and Middle Pleistocene deposits of the Kurkur Oasis. In *Desert and River in Nubia. Geomorphological and Prehistoric Environments at the Aswan Reservoir.* Butzer, K.W. and Hansen, C.L., with contributions by Leigh, E.G.J., Van Campo, M. and Gladfelter, B.G. (eds.) Madison: University of Wisconsin Press pp 515 – 520.

Van de Weerd, A. 1983 Palynology of some upper Miocene and Pliocene formations in Greece. *Geologisches Jahrbuch* 48: 3 – 63.

Van den Bergh, G.D., de Vos, J. and Sondaar, P.Y. 2001 The Late Quaternary palaeogeography of mammal evolution in the Indonesian Archipelago. *Palaeogeography, Palaeoclimatology, Palaeoecology,* 171: 385 – 408.

Van der Hammen, T. 1985 The Plio-Pleistocene climatic record of the tropical Andes. *Journal of the Geological Society (London)* 142: 483 – 489.

Vangengeim, E.A., Sotnikova, M.V., Alekseeva, L.I. and et al. 1988 *Biostratigrafiya pozdnego pliotsena-rannego pleistotsena Tadzhikistana. (Late Pliocene-Early Pleistocene Biostratigraphy of Tajikistan).* Moscow: Nauka.

Vekua, A. 1986 The Lower Pleistocene mammalian fauna of Akhalkalaki (Southern Georgia, USSR). *Palaeontographia Italica* 74 (12): 63 – 96.

Vekua, A. 1995 *Die Wirbeltierfauna des Villafranchium von Dmanisi und ihre biostratigraphische Bedeutung.* Mainz: Jahrbuch des Römisch-Germanischen Zentralmuseums Mainz.

Volkova, V.S. 1991 Pliocene climates of west Siberia. Pliocene climates of the Northern Hemisphere. *U.S. Geological Survey Open-File Report* 91 – 447: 44 – 45.

Vrba, E.S. 1980 The significance of bovid remains as indicators of environment and predation patterns. In *Fossils in the Making. Vertebrate Taphonomy and Paleoecology.* Behrensmeyer, A.K. and Hill, A.P. (eds.) Chicago: University of Chicago Press pp 247 – 271.

Vrba, E.S. 1992 Mammals as a key to evolutionary theory. *Journal of Mammalogy* 73: 1 – 28.

Vrba, E.S., Denton, G.H. and Prentice, M.L. 1988 – 1989 Climatic influences on early hominid behavior. *OSSA* 14: 127 – 156.

Walker, A.C. 1980 Functional anatomy and taphonomy. In *Fossils in the Making. Vertebrate Taphonomy and Paleoecology.* Behrensmeyer, A.K. and Hill, A.P. (eds.) Chicago: University of Chicago Press pp 182 – 196.

Wanpo, H. 1991 Quaternary cave accumulations and mammalian fauna in China. In *Quaternary Geology and Environment in China.* Liu Tungsheng (ed.) Beijing: Science Press pp 73 – 77.

Warren, R.E. 1990a Predictive modelling in archaeology: a primer. In *Interpreting Space: GIS and Archaeology.* Allen, K.M.S., Green, S.W. and Zubrow, E.B.W. (eds.) London: Taylor and Francis pp 90 – 111.

Warren, R.E. 1990b Predictive modelling of archaeological site location: a case study in the Midwest. In *Interpreting Space: GIS and Archaeology.* Allen, K.M.S., Green, S.W. and Zubrow, E.B.W. (eds.) London: Taylor and Francis pp 201 – 215.

Warren, R.E. and Asch, D.L. 2000 A predictive model of archaeological site location in the Eastern Prairie Peninsula. In *Practical Applications of GIS for Archaeologists. A Predictive Modeling Toolkit.* Wescott, K.L. and Brandon, R.J. (eds.) London: Taylor and Francis pp 5 – 32.

Webb, P.-N. and Harwood, D.M. 1993 Pliocene fossil *Nothofagus* (southern beech) from Antarctica: phytogeography, dispersal strategies, and survival in high latitude glacial-deglacial environments. In *Forest Development in Cold Climates.* Alden, J. and et al. (eds.) New York: Plenum Press pp 135 – 165.

Webster, P.J. 1987 The elementary monsoon. In *Monsoons.* Fein, J.S. and Stephens, P.L. (eds.) New York: J. Wiley & Sons, Inc. pp 3 – 32.

Wendorf, F. and Schild, R. 1980 *Prehistory of the Eastern Sahara.* London: Academic Press.

Wescott, K.L. and Brandon, R.J. (eds.) 2000 *Practical Applications of GIS for Archaeologists. A Predictive Modeling Toolkit.* London: Taylor and Francis

Wescott, K.L. and Kuiper, J.A. 2000 Using a GIS to model prehistoric site distributions in the Upper Chesapeake Bay. In *Practical Applications of GIS for Archaeologists. A Predictive Modeling Toolkit.* Wescott, K.L. and Brandon, R.J. (eds.) London: Taylor and Francis pp 59 – 72.

Western, D. 1980 Linking the ecology of past and present mammal communities. In *Fossils in the Making. Vertebrate Taphonomy and Paleoecology.* Behrensmeyer, A.K. and Hill, A.P. (eds.) Chicago: University of Chicago Press pp 41 – 54.

White, T.D. 1995 African omnivores: global climatic change and Plio-Pleistocene hominids and suids. In *Paleoclimate and Evolution with Emphasis on Human Origins.* Vrba, E.S., Denton, G.H., Partridge, T.C. and Burckle, L.H. (eds.) New Haven & London: Yale University Press pp 369 – 384.

Whybrow, P.J. 1984 Geological and faunal evidence from Arabia for mammal 'migrations' between Asia and Africa during the Miocene. *Courier Forschungsinstitut Senckenberg* 69: 189 – 198.

Whybrow, P.J., Hill, A., Yasin al-Tikriti, W. and Hailwood, E.A. 1990 Late Miocene primate fauna, flora and initial palaeomagnetic data from the Emirate of Abu Dhabi, United Arab Emirates. *Journal of Human Evolution* 19: 583 – 588.

Whybrow, P.J. and McClure, H.A. 1980/1981 Fossil mangrove roots and palaeoenvironments of the Miocene of the eastern Arabian peninsula. *Palaeogeography, Palaeoclimatology, Palaeoecology* 32: 213 – 225.

Whyte, R.O. (ed.) 1984 *The Evolution of the East Asian Environment. Palaeobotany, Palaeozoology and Palaeoanthropology.* Hong Kong: Centre of Asian Studies University of Hong Kong.

Wijninga, V.M. and Kuhry, P. 1990 A Pliocene flora from the Subachoque Valley (Cordillera Oriental, Colombia). *Review of Palaeobotany and Palynology* 62: 249 – 290.

Willard, D.A. 1992 Late Pliocene pollen assemblages from Ocean Drilling Project Hole 646B: census data and paleoclimatic estimates. *U.S. Geological Survey Open-File Report* 92 – 405: 12 pp.

Willard, D.A. 1994 Palynological record from the North Atlantic region at 3 Ma: vegetational distribution during a period of global warmth. *Review of Paleobotany and Palynology* 83: 275 – 297.

Willard, D.A., Cronin, T.M., Ishman, S.E. and Litwin, R.J. 1993 Terrestrial and marine records of climate and environmental change during the Pliocene in subtropical Florida. *Geology* 21: 679 – 682.

Williams, M., Dunkerley, D., De Deckker, P., Kershaw, P. and Chappell, J. 1998 *Quaternary Environments.* London: Arnold.

Williamson, P.G. 1985 Evidence for an early Plio-Pleistocene rainforest expansion in East Africa. *Nature* 315: 487 – 489.

Wise, S. 1998 *Research Training Programme GEO617. Introduction to Geographic Information Systems.* Sheffield: Department of Geography, University of Sheffield.

WoldeGabriel, G., White, T.D., Suwa, G., Renne, P., de Heinzelin, J., Hart, W.K. and Heiken, G. 1994 Ecological and temporal placement of Early Pliocene hominids at Aramis, Ethiopia. *Nature* 371: 330 – 333.

Wolfe, J.A. 1990 Estimates of Pliocene precipitation and temperature based on multivariate analysis of leaf physiognomy. In *Pliocene Climates: Scenario for Global Warming.* Open-File Report 90 – 64. Gosnell, L.B. and Poore, R.Z. (eds.) Washington: U.S. Geological Survey pp 39–42.

Wolsan, M. 1993 Évolution des carnivores Quaternaires en Europe Centrale dans leur contexte stratigraphique et paléoclimatique. *L' Anthropologie (Paris)* 97 (2/3): 203 – 222.

Wood, A., Whatley, R.C., Cronin, T.M. and Holtz, T. 1993 Pliocene paleotemperature recombination for the southern North Sea based on Ostracoda: a review. *Quaternary Science Reviews* 12 (9): 747 – 767.

Wood, B.A. and Turner, A. 1995 Out of Africa and into Asia. *Nature* 378: 239 – 240.

Wright, A., Fielding, A.H. and Wheater, C.P. 2000 Predicting the distribution of Eurasian badger (*Meles meles*) setts over an urbanized landscape: a GIS approach. *Photogrammetric Engineering and Remote Sensing* 66 (4): 423 – 428.

Wu Rukang, W.J.-K. and Xinzhi, W. 1984 Hominid fossils from China and their relation to those of neighbouring regions. In *The Evolution of the East Asian Environment. Palaeobotany, Palaeozoology and Palaeoanthropology.* Volume II. Whyte, R.O. (ed.) Hong Kong: Centre of Asian Studies University of Hong Kong pp 787 – 795.

Xiao Zhang, Arimoto, R. and Zhi An 1999 Glacial and interglacial patterns for Asian dust transport. *Quaternary Science Reviews* 18: 811 – 819.

Xu Ren 1981 Vegetational changes in the past and the uplift of the Qinghai-Xizang Plateau. In *Geological and Ecological Studies of Qinghai-Xizang (Tibet) Plateau.* Sinica, A. (ed.) Beijing & New York: Science Press, Gordon and Breach pp 139 – 144.

Xu Ren 1984 Changes of the vegetation in China since the Late Tertiary. In *The Evolution of the East Asian Environment. Palaeobotany, Palaeozoology and Palaeoanthropology.* Whyte, R.O. (ed.) Hong Kong: Centre of Asian Studies University of Hong Kong pp 426 – 432.

Xue Xiangxi and Zhang Yunxiang 1991 Quaternary mammalian fossils and fossil human beings in China. In *The Quaternary of China.* Zhang Zonghu (ed.) China: China Ocean Press pp 307 – 374.

Zagwijn, W.H. 1992 The beginning of the Ice Age in Europe and its major subdivisions. *Quaternary Science Reviews* 11: 583 – 591.

Zahlan, A.B. 1988a Societies and economies. Scientific and technical education. In *The Cambridge Encyclopedia of the Middle East and North Africa.* Mostyn, T. and Hourani, A. (eds.) Cambridge: Cambridge University Press pp 143 – 145.

Zahlan, A.B. 1988b Societies and economies. Scientific research. In *The Cambridge Encyclopedia of the Middle East and North Africa.* Mostyn, T. and Hourani, A. (eds.) Cambridge: Cambridge University Press pp 146 – 148.

Zalasiewicz, J.A., Mathers, S.J., Hughes, M.J., Gibbard, P.L., Peglar, S.M., Harland, R., Nicholson, R.A., Boulton, G.S., Cambridge, P. and Wealthall, G.P. 1988 Stratigraphy and palaeoenvironments of the Red Crag and Norwich Crag formations between Aldeburgh and Sizewell, Suffolk, England. *Philosophical Transactions of the Royal Society of London, B. Biological Sciences* 322: 221 – 272.

Zarate, M.A. and Fasana, J.L. 1989 The Plio-Pleistocene record of the central eastern Pampas, Buenos Aires, Province, Argentina: the Chapadmalal Case Study. *Palaeogeography, Palaeoclimatology, Palaeoecology* 72: 27 – 52.

Zazzo, A., Bocherens, H., Brunet, M., Beauvilain, A., Billiou, D., MacKaye, H.T., Vignaud, P. and Mariotti, A. 2000 Herbivore paleodiet and paleoenvironmental changes in Chad during the Pliocene using stable isotope ratios of tooth enamel carbonate. *Paleobiology* 26 (2): 294 – 309.

Zeitler, P.K. 1985 Cooling history of the NW Himalaya, Pakistan. *Tectonics* 4: 127 – 151.

Zeng Shaohua and Han Defen 1991 Quaternary mammals of China. In *Quaternary Geology and Environment in China.* Liu Tungsheng (ed.) Beijing: Science Press pp 101 – 114.

Zhang Zonghu (ed.) 1991 *The Quaternary of China.* China: China Ocean Press.

Zhanxiang Qiu and Zhuding Qiu 1995 Chronological sequence and subdivision of Chinese Neogene mammalian faunas. *Palaeogeography, Palaeoclimatology, Palaeoecology* 116 (1 – 2): 41 – 70.

Zhou, Z., Zhao, J. and Yin, P. 1989 Characteristics and tectonic evolution of the East China Sea. In *Chinese Sedimentary Basins.* Zhu, X. (ed.) Amsterdam: Elsevier Science pp 165 – 179.

Zhu, R.X., Hoffman, K.A., Potts, R., Deng, C.L., Pan, Y.X., Guo, B., Shl, C.D., Guo, Z.T., Yuan, B.Y., Hou, Y.M. and Huang, W.W. 2001 Earliest presence of humans in northeast Asia. *Nature* 413: 413 – 417.

Zickel, R.E. 1989 *Soviet Union: A Country Study.* http://lcweb2.loc.gov/frd/cs/sutoc.html.

www.ingramcontent.com/pod-product-compliance
Lightning Source LLC
LaVergne TN
LVHW071216210926
848482LV00015B/821